LITTLEWOOD'S MISCELLANY

The bust of Littlewood, by Gabriella Bollobás, in Trinity College.

LITTLEWOOD'S MISCELLANY

Edited by Béla Bollobás

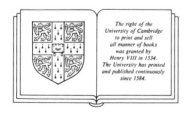

The right of the
University of Cambridge
to print and sell
all manner of books
was granted by
Henry VIII in 1534.
The University has printed
and published continuously
since 1584.

CAMBRIDGE UNIVERSITY PRESS
Cambridge
New York   New Rochelle
Melbourne   Sydney

Published by the Press Syndicate of the University of Cambridge
The Pitt Building, Trumpington Street, Cambridge CB2 1RP
32 East 57th Street, New York, NY 10022, USA
10 Stamford Road, Oakleigh, Melbourne 3166, Australia

First published in 1953 as '*A Mathematician's Miscellany*'
by Methuen & Co. Ltd., copyright 1953 Methuen & Co. Ltd.
The Mathematician's Art of Work,
© 1967 Rockefeller University Press, Illustrated by S. Wyatt.
This revised edition first published in 1986 by
Cambridge University Press,
© 1986 B. Bollobás
Reprinted 1988

Printed in Great Britain at the University Press, Cambridge

Library of Congress cataloguing in publication data available

British Library cataloguing in publication data available

ISBN 0 521 33058 0 hard covers
ISBN 0 521 33702 X paperback

CONTENTS

[†]Included in *A Mathematician's Miscellany*

To Ann

PREFACE

Acknowlegements are due for permission to reprint the following material. To the Editor of the *Mathematical Gazette* for 'Newton and the attraction of a sphere', 'Large numbers', and the review of Ramanujan's Collected Papers. To the Editor of the *Spectator* and Mr Arthur Robinson for the Competition Prize Entry, p. 58. To Rockefeller University Press for *The Mathematician's Art of Work*. To the President of the Dansk Matematisk Forening for permission to use the extract from Harald Bohr's *Collected Works*, pp. 8–11.

In writing *A Mathematician's Miscellany*, Professor Littlewood profited from criticisms and suggestions from Professor T. A. A. Broadbent, Dr T. M. Flett and Mr J. M. K. Vyvyan. Many of the diagrams in this volume were drawn from the originals of Dr Flett. Some additions to the original text are given in square brackets.

Special thanks are due to Mrs Ann Johannsen for her help and advice.

B. B.

FOREWORD

In 1951 the Mathematics Faculty in Cambridge asked J. E. Littlewood to give a talk, of about forty minutes, at the first of a series of 'social evenings'. A little later, the 'Archimedeans' — the Cambridge undergraduate mathematical society — invited him to address them. These two talks are the origins of a collection of essays for the general public which were published in 1953 as *A Mathematician's Miscellany.*

The volume was a great success. 'This admirable book is impossible to summarise. It overflows with what G. B. Shaw calls "the gaiety of genius" ' wrote one reviewer. 'For many of us this is the book of the year' added another. Several letters addressed to Littlewood began: 'What a delightful book.' From the many reactions to the book, Littlewood concluded that 'The loftier the intellect, the more the appreciation. The dim deprecate it.'

The *Miscellany* was reprinted several times but for the past twenty or so years it has been out of print. After writing the volume, Littlewood lived for another quarter of a century and went on collecting material for a new expanded edition: curiosities, howlers, strange anecdotes and various recollections of life at Trinity College. Much of that material is incorporated in this edition, together with the essay 'The Mathematician's Art of Work' which Littlewood wrote in 1967 and based on his collection of anecdotes.

The *Miscellany* remains a marvellous and attractive piece of work. However, the enjoyment and understanding of the reader will certainly be enhanced by knowing something of Littlewood's life and the environment at Trinity College in Cambridge, which formed the setting for so many of his stories. As in the original edition, a section marked by * is likely to be too technical for a non-mathematician.

The mathematical life in England in the first half of this century was dominated by two giants, Hardy and Littlewood. In the 1920s Edmund Landau, the eminent German number theorist, expressed the view that 'The mathematician Hardy-Littlewood was the best in the world, with Littlewood the more original genius and Hardy the better journalist.' To have deserved such praise is an extraordinary achievement at the best of times, but to appreciate it properly, we must recall that just a few years earlier England had no analysts to speak of. In the nineteenth century, France and Germany could boast of many outstanding pure mathematicians, and England, especially Cambridge, did have excellent applied mathematicians, including Green, Stokes, Adams, Lord Kelvin, Airy and Maxwell. However, in pure mathematics England produced only a handful of algebraists, among them Cayley and Sylvester, and failed to produce any notable analysts. This sorry state of affairs was changed by Littlewood and Hardy: by 1930 the school of analysis established by them was second to none.

John Edensor Littlewood was born at Rochester on 9 June 1885, the eldest of three children of Sylvia Maud Ackland and Edward Thornton Littlewood. His mother was part Irish, but his father was of pure British folk: farmers and landowners. Thorntons from the Eastern Counties, Robinsons of Suffolk, Stotherts of Scotland, Kitcheners of Binsteal and Littlewoods of Baildon Hall, Bradford. The father of the eminent journalist and diarist Henry Crabb Robinson was an ancestor of John Edensor Littlewood, and so was the great-grandfather of Lord Kitchener. It is recorded that a member of the family of Littlewood fought at Agincourt, and many branches of the family tree can be traced to the sixteenth century. This is not to say that Littlewood himself cared about his family tree: a mathematical proof containing gaps reminded him of being descended from William the Conqueror — with two gaps.

In recent centuries Littlewood's ancestors had been farmers, landowners, ministers, schoolmasters, printers, publishers, editors and doctors. Although Cambridge, according to Littlewood, inspired an awe equalled to nothing felt since, both his father and paternal grandfather were Cambridge men. The Reverend William Edensor Littlewood (1831–86) was educated at Pembroke College and was bracketed 35th wrangler in the mathematical tripos. This was the grandfather whose middle name, given in honour of his grandmother, Sarah Edensor, from the village of Edensor in Derbyshire, was passed on to John Edensor Littlewood. The eldest son of the Reverend Littlewood, Edward Thornton Littlewood (1859–1941), went to Peterhouse and was ninth wrangler. At the time College Fellowships were awarded on the basis of Tripos results and, but for a misplaced 'old school tie' attitude, he would have been Fellow of Magdalene. His College, Peterhouse, had no Fellowship to offer and he refused to take his parson father's advice and apply for one at

Magdalene, which went to a lower Wrangler. In his old age Littlewood remarked with a certain amount of sadness that his childhood would have been very different had his father stayed in Cambridge.

As it happened, Edward Littlewood accepted the headmastership of a newly founded school at Wynberg, near Cape Town, and took his family there in 1892. J. E. Littlewood spent eight years of his childhood in South Africa, the beauty of which made an impression on him that he never forgot. Having grown out of the schools there, he went to the Cape University, but his father realised that his mathematical education would suffer if he stayed in South Africa, so in 1900 he was sent to St. Paul's School in London. He spent almost three years at the school, under the guidance of F. S. Macaulay, an unusually able mathematician, who in 1928 became a Fellow of the Royal Society. (Not surprisingly, it was Littlewood who proposed him.) Littlewood's work at the school and his subsequent life at Cambridge are admirably related in *A Mathematical Education*, so I will not dwell on the details.

Littlewood took the Entrance Scholarship Examination of December 1902 and although he was expected to do well, he found the papers too difficult and got only a minor scholarship at Trinity College.

On arriving in Cambridge, Littlewood began work for the *Mathematical Tripos*. In its prime the Mathematical Tripos was far and away the most severe mathematical test that the world has ever known, one to which no university today (including Cambridge) can show any parallel. The Examination evolved during the eighteenth century; from 1753 on the candidates were divided into three classes: Wranglers, Senior Optimes and Optimes. In order to establish a *strict order of merit*, the examination was turned into a high-speed marathon: four days of tests of up to ten hours a day. This absurd examination produced few excellent pure mathematicians, but it was phenomenally successful in training outstanding applied mathematicians.

The candidates in the Tripos worked under various *coaches*, who drilled their men mercilessly. In spite of the name, a good coach was usually a respectable mathematician and, occasionally, a very good one. Littlewood was lucky to have the last of the great coaches, R. A. Herman, who was a contemporary and a friend of his father, and a Fellow of Trinity. To be in the running for Senior Wrangler, the top man in the order of merit, undergraduates had to spend two-thirds of the time practising how to solve difficult problems against time, and this is what Littlewood did.

In Littlewood's time Part I of the Mathematical Tripos was a three-year course but occasionally scholars used to take Part I at the end of their second year. Littlewood did this in 1905, while still 19, and was bracketed Senior Wrangler with Mercer, who had graduated from Manchester University before coming to Cambridge.

The Senior Wranglers were celebrities in Cambridge and their photos were sold during May Week. When a friend of his tried to buy one of him, he was told: 'I'm afraid we're sold out of Mr Littlewood but we have plenty of Mr Mercer.'

Littlewood as Senior Wrangler.

A few years later, in 1910, Hardy played a decisive role in abolishing the strict order of merit in the Tripos, and was a sworn enemy of the milder examination which replaced it. Littlewood was also firmly against the order of merit; he thought that his first two years at Cambridge were wasted although he did not feel that the system caused him any real harm. As Hardy wrote later: 'He understood that the mathematics he was studying was not the real thing and regarded himself as playing a game. It was not exactly the game he would have chosen, but it was the game which the regulations prescribed, and it seemed to him that, if you were going to play the game at all, you might as well accept the situation and play it with all your force. He believed that he could play the game as well as any of his rivals, and he was right.' He even felt a satisfaction of a sort in successful craftsmanship.

Littlewood took Part II of the Mathematical Tripos in his third year. Although he was learning genuine mathematics he wasted a good deal of time in the ordinary course of trial and error. At the end of the third year, in the Long Vacation of 1906, Littlewood began research under E. W. Barnes (later Bishop of Birmingham). As the first project, Barnes suggested study of entire functions of order zero. After a few months Littlewood's efforts resulted in a fifty page paper.

Encouraged by Littlewood's success, Barnes suggested another problem: 'Prove the Riemann Hypothesis.'

The *Riemann Hypothesis* (R.H.) is, by general concensus, the most important unsolved problem in Pure Mathematics. *The zeta function of Riemann is defined for $s = \sigma + it$, σ and t real, $\sigma > 1$, by

$$\varsigma(s) = \frac{1}{1^s} + \frac{1}{2^s} + \frac{1}{3^s} + \cdots$$

This function is regular in the half-plane $\sigma > 1$ and it has an analytic continuation throughout the s plane, having a simple pole at $s = 1$. At first sight $\zeta(s)$ is a peculiarly defined complex function but, in fact, it is intimately related to the distribution of primes. Indeed, the great eighteenth century mathematician Leonhard Euler knew that

$$\varsigma(s) = \prod_p \left(1 + \frac{1}{p^s} + \frac{1}{p^{2s}} + \frac{1}{p^{3s}} + \cdots\right)$$

where the product is taken over all primes.

The main questions about $\varsigma(s)$ concern the distribution of its zeros. It is known that every negative even integer is a zero of $\varsigma(s)$ (these are the *trivial zeros*) and that infinitely many zeros lie in the *critical strip* $0 < \sigma < 1$. In 1860, the outstanding German mathematician Riemann conjectured that all non-trivial zeros lie on the *critical line* $\sigma = 1/2$. This is the Riemann Hypothesis, which is still open today. In terms of the distribution of prime numbers, R.H. means that the primes are fairly regularly distributed.

The *Prime Number Theorem* (P.N.T.), proved independently by Hadamard and de la Vallée Poussin, asserts that $\pi(x)$, the number of primes up to x, is about $x/\log x$, and the *logarithmic integral* $li(x) = \int_0^x \frac{dt}{\log t}$ is an even better approximation. (The integral is defined by its *Cauchy principal value*:

$$li(x) = \lim_{\epsilon \to 0} \left\{ \int_0^{1-\epsilon} + \int_{1+\epsilon}^x \right\}.)$$

Assuming R.H., the P.N.T. can be improved to

$$|\pi(x) - li(x)| \leq C x^{1/2} \log x,$$

where C is some absolute constant. (In one of his most celebrated papers, Littlewood later proved that, contrary to all the numerical evidence, the difference $\pi(x) - li(x)$ changes sign infinitely often, (see page 100)*

Barnes did not know that R.H. was connected to the distribution of primes, although that had been proved on the continent several years before, and Littlewood had to discover it for himself: assuming R.H., he deduced the Prime Number Theorem. This was just in time for his first Fellowship dissertation.

Trinity offers annually a number of junior fellowships to its graduates, who have three opportunities of competing: at the end of the fourth, fifth and sixth years following their matriculation. Littlewood competed at his first opportunity, in September 1907. The dissertation was well received and would have secured his election if there had not been a candidate in classics competing at his last chance whom the electors considered to deserve election. Littlewood was informed that his election in the following year, 1908, was a certainty, and the election duly took place. The first paper which made Littlewood famous, published in 1912, was also about some consequences of the Riemann hypothesis.

Meanwhile Littlewood had been offered the Richardson lectureship in the University of Manchester. Though at £250 this was better than the usual £150 or £120, he did not gain financially, but felt he needed a change from Cambridge. On looking back he considered that it was a disaster on his part to accept it, for he was greatly overworked during the three years of his tenure. He always spoke of it as his period of *exile*. Once, during his period of exile, he walked along a river in Manchester, and it looked like ink. Presently a tributary ran into it, making an inky trace on the surface. King John sprang to his mind: 'Hell darkened as he entered it.'

Littlewood joined the Trinity staff in 1910, replacing Whitehead. This coincided with new mathematical interests. Landau's fundamental book on analytical number theory had been published only a year earlier,

Littlewood in 1907.

which enabled Hardy and Littlewood to catch up with the latest results in number theory and to confront the analytical problems they give rise to. In 1920 Littlewood succeeded Hardy to the Cayley lectureship in the University.

In 1912 he moved into a large set of rooms on the first floor of Nevile's Court. He occupied these for the next 65 years, until his death, except during the First World War, when he served as Second Lieutenant in the Royal Garrison Artillery. Throughout all these years, he was a much loved and respected Fellow of Trinity. He felt perfectly at home in the College and was deeply attached to it. He never cared for College office but nonetheless he played a key role in the society and, for several decades, shaped its development by serving as a Fellowship Elector.

Shortly before the War, Hardy and Littlewood began their extraordinarily successful collaboration, lasting for 35 years — surely the most successful collaboration ever in mathematics! They wrote a hundred joint papers, with their last publication being published a year after Hardy's death. In addition, with Pólya, they wrote an excellent book entitled *Inequalities*, published by CUP in 1934, which is widely used to this day.

There are many reasons why the Hardy-Littlewood collaboration flourished. They had a number of common interests, especially summability, inequalities, Diophantine approximation and its connections to function theory, Fourier series and the theory of numbers, inspired by Landau's *Primzahlen*. They were both geniuses, completely dedicated to mathematics. Hardy was, perhaps, more stylish, a man of intellectual panache, interested in beautiful patterns, but Littlewood was imaginative and amazingly powerful, enjoying the challenge of a very difficult problem.

This period is vividly described by the eminent Danish mathematician Harald Bohr, at a lecture given on his sixtieth birthday in 1947 (*Collected Works*, vol. I, pp. *xxvii – xxviii*, 1953, Dansk Mat. Forening, reproduced with permission).

> Already early in life, I had the good fortune to come into close professional contact — which later turned into an intimate friendship — with the two only slightly older English mathematicians, Hardy and Littlewood, who were to bring English pure mathematics to such a high standard. Thus I often had occasion to take a trip to Cambridge, the classical centre of English mathematics and natural sciences since the days of Newton, and the old university town of Oxford, with which Hardy was connected for some years. Life in the old English university colleges — for me it was Trinity College in Cambridge and New College in Oxford — could not but captivate and enchant everyone. While everything was permeated and marked by venerable traditions, unbroken through centuries, at the same time there reigned a rare spirit of freedom and tolerance, and not only was it allowed, but it was even appreciated that even the most individual and divergent opinions were expressed undisguisedly, often in extreme form, though never in an offensive manner.
>
> To illustrate to what extent Hardy and Littlewood in the course of the years came to be considered as the leaders of recent English mathematical research, I may report what an excellent colleague once jokingly said: 'Nowadays, there are only three really great English mathematicians: Hardy, Littlewood and

Stowford House,
Bideford

Dear Hardy,

On skimming Landau Bd. 2
I gather from p.871 that from
$\{\omega\}<t^{\varepsilon}$ $|\zeta(\sigma+iz)| \asymp z^{\varepsilon}$ follows

$\sum \frac{d(n)}{n^s}$ is cgt. for $\sigma > \frac{1}{2}$. I don't
remember what $p(n)$ is but I see he (our)
has deduced from the Riemann assumption
that the series cgs for $\sigma > .83$.

I find that $\left(\frac{d}{dx}\right)^n \sum a_n x^n \to$ limit
for all x, or $|a_n| < K n^{-\alpha}$ $(\alpha > 0)$ do
not involve $\sum a_n$ cgt. In fact nothing
more seems to come from the fact that
(than my Tauber)

A letter from Littlewood to Hardy, c.1910

Hardy-Littlewood.' The last refers to the marvellous collaboration through the years between these two equally outstanding
scientists with their very different personalities. This cooper-
ation was to lead to such great results and to the creation of
entirely new methods, not least in the theory of numbers, that
to the uninitiated, they almost seemed to have fused into one.
To illustrate the strong feelings of independence which, as a
part of the old traditions, are so characteristic of the English
spirit, I should like to tell how Hardy and Littlewood, when
they planned and began their far-reaching and intensive team
work, still had some misgivings about it because they feared

Hardy and Littlewood in New Court, Trinity College.

that it might encroach on their personal freedom, so vitally important to them. Therefore, as a safety measure, (it was, as usual when they worked out something together, Hardy who did the writing), they amused themselves by formulating some so-called 'axioms' for their mutual collaboration. There were in all four such axioms. The first of them said that, when one wrote to the other (they often preferred to exchange thoughts in writing instead of orally), it was completely indifferent whether what they wrote was right or wrong. As Hardy put it, otherwise they could not write completely as they pleased, but would have to feel a certain responsibility thereby. The second axiom was to the effect that, when one received a letter from the other, he was under no obligation whatsoever to read it, let alone to answer it, — because, as they said, it might be that the recipient of the letter would prefer not to work at that particular time, or perhaps that he was just then interested in other problems. And they really observed this axiom to the fullest extent. When Hardy once stayed with me in Copenhagen, thick mathematical letters arrived daily from Littlewood, who was obviously very much in the mood for work, and I have seen Hardy calmly throw the letters into a corner of the room, saying: 'I suppose I shall want to read them some day.' The third axiom was to the effect that, although it did not really matter if they both thought about the same detail, still, it was preferable that they should not do so. And, finally, the fourth, and perhaps most important axiom, stated that it was quite indifferent if one of them had not contributed the least bit to the contents of a paper under their common name; otherwise there would constantly arise quarrels and difficulties in that now one, and now the other, would oppose being named co-author. I think one may safely say that seldom — or never — was such an important and harmonious collaboration founded on such apparently negative axioms.

In 1914 the Royal Society Council decided that the proposer of a candidate should renew the form *yearly*. In 1915, when this came into operation, E. W. Hobson naturally forgot. As a result, when they wanted to consider Littlewood, it was constitutionally impossible. The idea was at once dropped but Littlewood had to wait till the next year.

In January 1913, Hardy received an unsolicited letter from Madras. The writer was a certain S. Ramanujan, a clerk in the Port Trust Office, a man of 23, without any formal mathematical training. Ramanujan sent Hardy some of his theorems and asked him to publish anything of value. It did not take long for Hardy and Littlewood to conclude that Ramanujan was a man of exceptional ability, and they decided to bring Ramanujan to Cambridge. As usual, Trinity College was generous in supporting genuine research: Ramanujan was offered a scholarship and duly arrived in Cambridge in April 1914. Hardy was soon convinced that, in terms of natural talent, Ramanujan was in the class of Euler and Gauss. Sadly, in May 1917, Ramanujan fell ill, returned to India in 1919 and died in 1920.

Although Littlewood did no joint work with Ramanujan, which is not very surprising since he was away from Cambridge during most of Ramanujan's time there, he did benefit from Ramanujan's stay. He also helped Hardy to get Ramanujan elected a Fellow of the Royal Society and was instrumental in securing him a Fellowship in Trinity College.

Trinity College,
Cambridge.

30.5.29

A note from Littlewood to Hardy.

In 1920 Hardy was elected Savilian Professor of Mathematics in Oxford and left Cambridge, returning to Trinity College only eleven years later. The Hardy-Littlewood collaboration continued unabated during these eleven years, resulting in the famous *Partitio Numerorum* papers, many of the papers on the theory of series and the famous and important maximal theorem. Littlewood's own work during the 1920s was largely concentrated on complex function theory.

Littlewood's long deserved professorship materialised in 1928, when he became the first occupant of a chair of mathematics established under the will of W. W. Rouse Ball, a Fellow of Trinity. The appointment would have given the founder particular pleasure, for Littlewood had been one of his favourite pupils. At the final stage, when things were settled, Hardy delivered a ten-minute speech he prepared very carefully, ending: 'It may be allowed to Littlewood's best pupil to move the formal motion.' In the following year Littlewood was awarded a Royal Medal of the Royal Society.

In December 1931, Hardy and Littlewood announced weekly meetings of a conversation class to start in January 1932 in Littlewood's rooms. According to E. C. Titchmarsh, 'this was a model of what such a thing should be. Mathematicians of all nationalities and ages were encouraged to hold forth on their own work, and the whole exercise was conducted with a delightful informality that gave ample scope for free discussion after each paper.' Nevertheless, as Dame Mary Cartwright wrote, a little later there was a metamorphosis of Littlewood's conversation class into a larger gathering run by Hardy.

Although Littlewood's main collaborator was Hardy, he collaborated with many other excellent mathematicians; including R. E. A. C. Paley, his best student, in an important series of papers on Fourier series, with A. C. Offord on the distribution of the zeros and α values of random integral functions, and with Dame Mary Cartwright on non-linear differential equations. Paley, born in 1907, was one of the greatest stars in pure mathematics in Britain, whose young genius frightened even Hardy. Had he lived, he might well have turned into another Littlewood: his 26 papers, written mostly in collaboration with Littlewood, Zygmund, Wiener and Ursell, opened new areas in analysis. Alas, he died in an avalanche while skiing in Banff, Alberta at the tragically early age of 26.

This is what Littlewood wrote about his collaboration with Mary Cartwright:

'Two rats fell into a can of milk. After swimming for a time one of them realised his hopeless fate and drowned. The other persisted, and at last the milk was turned to butter and he could get out.

'In the first part of the war, Miss Cartwright and I got drawn into van der Pol's equation. For something to do we went on and on at the thing with no earthly prospect of "results": suddenly the entire vista of the dramatic fine structure of solutions stared us in the face.'

In 1950, at the statutory age of 65, Littlewood retired and became an Emeritus Professor. The Faculty Board realised that it would be madness to lose the services of the most eminent mathematician in England, so they wrote to the General Board:

'Professor Littlewood is not only exceptionally eminent, but is still at the height of his powers. The loss of his teaching would be irreparable,

Littlewood lecturing in Cambridge.

and it is avoidable. Permission is requested to pay a fee of the order of £100 for each term's course of lectures.'

The response: £15, the fee to an apprentice giving his first course as a try out, to a class of 2 or 3.

So Littlewood gave courses at £15 for 4 years. He tried to stop once but there was a cry of distress. At the same time he turned down lucrative offers from the United States.

Littlewood liked the sea from his student days, when he spent most of his holidays at Bideford, where his uncle had his medical practice. Later he spent many of his vacations away from Cambridge along the Cornish coast where he used to swim. For several summers Bertrand Russell, Hardy and Littlewood vacationed together, often discussing general relativity which was then comparatively new. Later he shared a house at Treen with the Streatfeild family, friends of long standing. and he adopted the Streatfeild children, who called him 'Uncle John'. In Cornwall he took up rock climbing, attaining a high standard in the art.

From about 1930 on, his happiness was considerably marred by an obscure nervous malady which afflicted him for about thirty years. It was impossible for anyone not intimate with him to realise that he was a sick man. Even the eccentric Russian émigré, A. S. Besicovitch, who was a close colleague and succeeded Littlewood as Rouse Ball professor, refused to take his illness seriously. For his lectures were throughout of the highest calibre, he took more than his share of research pupils, he performed admirably year after year the arduous task of Fellowship Elector in the annual Trinity Research Fellowship Competition and he continued to be creative at the highest level. That his research was of the first order both in quality and quantity, is attested by two Royal Society awards: the Sylvester medal in 1943 and the Copley medal in 1958. Nevertheless, Littlewood was adamant in his assertion that for about thirty years he was functioning at half his capacity, often spending hours in cinemas just to while away the tedious hours.

He took up skiing in Switzerland in 1924 and soon acquired a high degree of proficiency. After experimenting for some years with several places, in 1935 he settled on the Hotel Meierhof in Davos, which he visited every year, except during the war, accompanied by Ann Streatfeild, whom he called his niece.

The year 1960 was a memorable one for him. A severe attack of influenza led him to an able Trinity doctor, Edward Bevan, who, on learning of his depression, sent him to a brilliant neurologist, Dr Beresford Davies. The latter, after experimenting with various drugs which were then comparatively new, hit upon the right combination for Littlewood's case, and cured him. His freedom from depression led him to accept invitations to the United States, which he would never have contemplated during his illness. He made eight visits, holding visiting professorships at the Universities of Michigan, Wisconsin (Madison and Milwaukee), Chicago, California (Berkeley) and at the Rockefeller University. During one of these visits he made an extensive lecture tour, which had been organized for him by Besicovitch who was then living in America. It must be unique for a man who had never lived in the USA before the age of 75 to make so many professional visits there.

Littlewood remained active in mathematics even at an advanced age: his last paper was published in 1972, when he was 87. One of his most intricate papers, concerning Van der Pol's equation and its generalisations, was written when he was over seventy: 110 pages of hard analysis, based on his joint work with Mary Cartwright. He called the paper 'The Monster' and he himself said of it: 'It is very heavy going and I should never have read it had I not written it myself.' His last hard paper, breaking new ground, was published in the first issue of the Advances in Applied Probability, when he was 84. In this paper he gave very precise bounds for the probability in the tail of the binomial distribution. The bounds he gave are still the best: in fact, they are so fine that their precise form is rarely needed; Littlewood himself became interested in these estimates because he conducted long 'card guessing' experiments with Ann.

In August 1977 Littlewood fell out of bed during the night. He could neither get back to bed nor call for help. When he was found in the morning he was very unwell. He was taken to the Evelyn Nursing Home in Cambridge, where he died on 6 September 1977.

Throughout his life Littlewood was fascinated by the Riemann Hypothesis, and he proved several related results which, after 50 years or so, are still the best known. A problem as famous as this attracts many mathematicians and spurious proofs are produced frequently.

In 1971 a paper was submitted to the London Mathematical Society, claiming a proof of R.H. The author sent a copy to Littlewood as well. For a while Littlewood was patient. But months went by and the editors still held on to the paper. Was it perhaps correct? Littlewood became more and more demanding so I hawked the paper around the mathematics department and even wrote to the editors of the Journal. I had no luck: nobody was willing to waste his time trying to penetrate the complicated (and somewhat old fashioned) formulae. Eventually I had to give in to Littlewood's demand and we began to read the paper together. I knew nothing about the formulae and I was amazed how well *he* knew them. After a few hours of painstaking work he was relieved to find a mistake. But that isn't the end of the story. A week or so later I got the referee's report, pointing out a mistake which was *not* there.

Although Littlewood knew that the first 60,000 zeros of the zeta function were on the critical line, he became increasingly sceptical of R. H., saying that there was no convincing reason why the result should be true. In spite of this, he was infatuated with the problem and often told a story proving that he was not alone in that.

There is a German legend about Barbarossa, the emperor Frederick I. The common people of Germany liked him and as he died in a crusade and was buried in a far away grave, the legend sprang up that he was still alive, asleep in a cavern but would wake and come out, even after hundreds of years, when Germany needed him.

Somebody allegedly asked the famous German mathematician, David Hilbert, 'If you were to revive, like Barbarossa, after five hundred years, what would you do?' 'I would ask,' said Hilbert, 'Has somebody proved the Riemann hypothesis?'

In 1972 Littlewood had two bad falls and he fell again in January 1975. He was taken to the Evelyn Nursing Home in Cambridge, but he had very little interest in life. In my desparation I suggested the problem of determining the best constant in Burkholder's weak L_1 inequality (an extension of an inequality Littlewood had worked on). To my immense relief (and amazement), Littlewood became interested in the problem. He had never heard of martingales but he was keen to learn about them so he was happy to listen to my brief explanations and was willing to read some introductory chapters! All this at the age of 89 and in bad health! It seemed that mathematics did help to revive his spirits and he could leave the nursing home a few weeks later. From then on, Littlewood kept up his interest in the weak inequality and worked hard to find suitable constructions to complement an improved upper bound. Unfortunately, we did not have much success so eventually I published the improvement only after Littlewood's death.

In the early seventies Littlewood became more and more concerned about his 'niece', Ann Streatfeild. He frequently confided these concerns to my wife Gabriella when she accompanied him on long walks in the Fellows' Garden or sat with him in his rooms. Many a peaceful morning was spent by Gabriella sculpting him (several busts were made and destroyed later), while they listened to Bach, Beethoven or Mozart. He considered life too short to waste on other composers. In the afternoon Littlewood drank a large glass of vodka diluted with water; a habit he acquired from Besicovitch. Gabriella often read to him from Omar Khayyam, the Persian mathematician, who was his favourite poet, and then he would meditate on the immortal lines while the music poured over him. He was particularly fond of the 28th quatrain of the Rubaiyat:

> With them the Seed of Wisdom
> did I sow,
> And with my own hand
> labour'd it to grow:
> And this was all the
> Harvest that I reap'd–
> 'I came like Water, and like
> Wind I go.'

Ann was an excellent skier but she was not married and Littlewood

TRINITY COLLEGE
CAMBRIDGE
CB2 ITQ

Dear Bela,

I am so sorry, but the proof of √c has mysteriously disappeared, & I can't reconstruct the "vital step". Would you be so kind as to send it again?

yours
Jack Littlewood

A letter from Littlewood in 1975.

was worried what would happen to her after his death. This was not too surprising, for Ann was actually his daughter, though he kept it a secret. However, he was bothered about this secrecy. Eventually Gabriella and I persuaded him to refer to Ann as his daughter. One night in the Combination room he began to speak of 'my daughter' without offering any explanation. Next day he was most depressed that nobody had blinked an eyelid! Not a single question was asked. When Ann married Carl Johannsen of Zurich, Littlewood was very happy indeed. She would

A champagne dinner with Ann in Davos.

not be left alone, after all.

Ann saw Littlewood as her affectionate 'Uncle John' and, with the short Cambridge terms, he spent much of his time and did much of his mathematics in her company. When his daughter was young, Littlewood wrote once a week to her, often enclosing cowslips, which he had picked and pressed, or puzzles and stamps to interest her. Ann has fond memories of their summers in Cornwall.

> He had a rigid timetable in Treen. Coffee in bed. Work, when possible, in the sun on our porch, where he had a broken-down chair, a log to put up his feet on and stones to weigh down his papers. He then went for a swim, timing himself exactly. In the early days he would go for another swim in the afternoon. If the weather was really too bad for swimming (and it had to be *very* bad), he had his 'walks' of various lengths. After his China tea he played Patience until it was time to fetch his beer from the pub, which he drank with his evening meal. Afterwards we often played cards or other games.

> Despite his liking of a routine, he would always be ready to upset it for me if, for instance, I had time to go swimming or climbing with him. He was a strong swimmer: he could swim almost indefinitely on his back which, he maintains, was the most efficient way. We used to go for long swims together,

when we swam for half an hour or more.

He did a lot of rock climbing alone, but he loved to organise expeditions for anyone he could rope in. They set off with ropes and rucksacks with a picnic lunch. My mother went too in the early days.

Uncle John gave up climbing in the war and was never so confident on rock again. When he was 80 I took him for the last time on the rope to 'our' beach by 'our' climb. (It's not at all difficult, but plenty of young people have turned pale and got stuck on it.)

Their holidays together at Davos in Switzerland also bring back happy recollections.

Uncle John didn't ski much either after the war. He had lost over six years and never mastered the 'new' technique. Uncle John and I settled into a routine. I used to write all his letters from Switzerland, at first giving them to him to sign. But even signing became a trouble for him, and I learned to forge his signature very successfully.

He went for long walks and on Sundays I walked with him. We had afternoon tea together in his hotel room. It was after tea that I had my 'lectures'. After dinner, we played two parties of Piquet, keeping the scores over the years in a series of little note books.

Littlewood was an incredibly impressive man in every respect. He was as unlike as possible to what used to be and still is regarded as a typical mathematician. He was short but very strong and a great sportsman: he was one of the best gymnasts at school, loved to swim, stroked a College eight at Cambridge, excelled at rock climbing and skiing, was a hard hitting cricketer (though he did not play regularly). He was an elegant dancer and passionately keen on music. He was well read and could talk on almost everything in an engaging fashion. When I got to know him, in 1969, he had already been the Senior Fellow of Trinity College for seven years. He was admired and revered by all the Fellows. After dinner in Hall, he invariably went up to the Combination Room to have port or claret. Littlewood refused to preside, but his presence greatly enhanced the atmosphere of the evening. He was one of the few Fellows to take snuff and after a good sneeze he always remarked: 'When you sneeze, you are closest to death' — a reference to the Black Death. It was rumoured that he had a knack of choosing a place where he would get a 'buzz': an extra glass for having finished the bottle. When in the early seventies he thought that he could no longer afford his port in the Combination Room, the College Council decided that the Senior Fellow should have his drink free, whether he presided or not.

In addition to the senior Cambridge mathematicians: Besicovitch, Charles Burkill, Mary Cartwright and Louis Mordell, Littlewood was particularly friendly with Harry Hollond and Michael Vyvyan. Hollond was a Law Fellow who had first come to the college at the same time as Littlewood. They both had a deep interest in music, especially Bach, Beethoven and Mozart. Vyvyan is a History Fellow who shared Littlewood's interest in climbing.

His doctor and friend, Edward Bevan, dropped in on him several times a week, to check whether he needed any change in the drugs he was taking or whether he had caught cold.

Littlewood wearing his favourite silk jacket.

Littlewood was very witty, had a lovely sense of humour and was full of fascinating stories. In his younger days, incompetent research students reminded him of a general in *War and Peace* who, disturbed by the groans of a soldier, calls to him: 'My good man, do try to die more quietly!' He quoted Bismarck: 'The English summer is the winter painted green.' When Gabriella told him that a mathematician left Trinity College to be the Master of another College, he exclaimed: 'What a come down!' (No doubt he remembered Clemenceau who, on hearing that Paderewski was President of the Polish Republic, asked: 'Is that the pianist?' 'Yes.' 'What a come down!')

A story which Littlewood used to good effect was about a young man, painting a very tough old man: 'I should enjoy still more painting you when you are 100.' 'I don't see why you shouldn't, you look pretty healthy.'

Another of his stories concerned his parents, who moved to Cambridge in 1920 upon the retirement of his father from the headmastership at Wynberg. 'My Father and Mother delighted in University Functions, and also in any kudos of mine. At an Honorary degree lunch (myself absent) my Mother was put between Eddington and Hopkins (then President of the Royal Society); both played up to the limit. Hollond, opposite, volunteered that the Trinity Council had resolved in this *one instance* to waive the rule that only *one* relation of a Fellow should be invited and Hopkins weighed in shamelessly "quite right too, for so great a man".'

I can do no better than finish this introduction with the words which Sir Henry Dale used in 1943 to summarise Littlewood's achievements for the Royal Society:

> Littlewood, on Hardy's own estimate, is the finest mathematician he has ever known. He was the man most likely to storm and smash a really deep and formidable problem; there was no one else who could command such a combination of insight, technique and power.

B. B.

INTRODUCTION TO
A MATHEMATICIAN'S MISCELLANY

§1. A Miscellany is a collection without a natural ordering relation; I shall not attempt a spurious unity by imposing artificial ones. I hope that variety may compensate for this lack, except for those irreconcilable persons who demand an appearance of unity and uniform level.

Anyone open to the idea of looking through a popular book on mathematics should be able to get on with this one. I will describe, and sometimes address, him as an 'amateur'. I constantly meet people who are doubtful, generally without due reason, about their potential capacity. The first test is whether you got anything out of geometry. To have disliked or failed to get on with other subjects need mean nothing; much drill and drudgery is unavoidable before they can get started, and bad teaching can make them unintelligible even to a born mathematician. If your education just included, or just stopped short of including, 'a little calculus', you are fairly high in the amateur class.

The book contains pieces of technical mathematics, on occasion pieces that only a professional mathematician can follow; these have been included as contributing to the full picture of the moment as viewed by the professional, but they can all be skipped without prejudice to the rest, and a coherent story will remain. I have enclosed between *'s sections which the amateur should probably skip (but he need not give up too soon). Outside these I have aimed consciously at a level to suit his needs (and here it is the professional who will have to skip at times).

The qualities I have aimed at in selecting material are two. First relative unfamiliarity, even to some mathematicians. This is why some things receive only bare mention. They complete a picture (like the technical pieces above) but are not essential to the amateur (anything that is is given in full). He should on no account be put off if he does not happen to know them (and I generally give references). A specific case happens at the very beginning; (1) of the next chapter and the succeeding paragraph. 'Familiar' here means 'familiar to the mathematician'. But experience shows that some amateurs will know Euclid's proof; if so, they will also know that it is so familiar that I must not discuss it[1]; it shows on the other hand that some do not know it; it is they who must not be put off.

The other quality is lightness, notwithstanding the highbrow pieces; my aim is entertainment and there will be no uplift. I must leave this to the judgment of my readers, but I shall have failed where they find anything cheap or trivial. A good mathematical joke is better, and better mathematics, than a dozen mediocre papers.

---

[1]There is, however, a one line indication for the professional on p. 40

MATHEMATICS WITH MINIMUM 'RAW MATERIAL'

§1. What pieces of genuine mathematics come under this? A *sine qua non* is certainly that the *result* should be intelligible to the amateur. We need not insist that the *proof* also should (see e.g. examples (15), (16), (19)), though most often it is. I begin with clear cases; later on we shade off and the latest examples are there because for various reasons they happen to appeal to my particular taste. Various things that belong to a complete picture are 'mentioned' or postponed.

(1) Euclid's ('familiar') proof that the primes are infinite in number is obviously in the running for the highest place. (See e.g. Hardy's *A Mathematician's Apology*, pp. 32–34, or p. 40 below.) The 'familiar' things in §3 belong here; but to make them intelligible to the amateur would call for interrupting explanation.

My actual choice for first place is a well-known puzzle that swept Europe a good many years ago and in one form or another has appeared in a number of books. I revert to the original form, in which A's flash of insight is accounted for by an emotional stimulus.

(2) Three ladies, A, B, C in a railway carriage all have dirty faces and are all laughing. It suddenly flashes on A: why doesn't B realize C is laughing at her? — Heavens! *I* must be laughable. (Formally: if I, A, am not laughable, B will be arguing: if I, B, am not laughable, C has nothing to laugh at. Since B does not so argue, I, A, must be laughable.)

This is genuine mathematical reasoning, and surely with minimum material. But further, which has not got into the books so far as I know, there is an extension, in principle, to n ladies, all dirty and all laughing. There is an induction: in the $(n + 1)$-situation A argues: if I am not laughable, $B, C, \ldots$ constitute an n-situation and B would stop laughing, but does not.

Compare the rule for toasting 3 slices of bread on a toaster that holds only 2. A_1, B_1; then B_2, C_1; then C_2, A_2. This falls short of being mathematics.

(3) The following will probably not stand up to close analysis, but given a little goodwill is entertaining.

There is an indefinite supply of cards marked 1 and 2 on opposite sides, and of cards marked 2 and 3, 3 and 4, and so on. A card is drawn at random by a referee and held between the players A, B so that each sees one side only. Either player may veto the round, but if it is played the player seeing the higher number wins. The point now is that every round is vetoed. If A sees a 1 the other side is 2 and he must veto. If he sees a 2 the other side is 1 or 3; if 1 then B must veto; if he does not then A must. And so on by induction.

(4) An analogous example (Schrödinger) is as follows. We have cards similar to those in (3), but this time there are 10^n of the ones of type $(n, n + 1)$, and the player seeing the *lower* number wins. A and B may now bet each with a bookie (or for that matter with each other), backing themselves at evens. The position now is that whatever A sees he 'should' bet, and the same is true of B, the odds in favour being 9 to 1. Once the monstrous hypothesis has been got across (as it generally has), then, whatever number n A sees, it is 10 times more probable that the other side is $n + 1$ than that it is $n - 1$. (Incidentally, whatever number N is assigned before a card is drawn, it is infinitely probable that the numbers on the card will be greater than N.)

(5) *An infinity paradox.* Balls numbered $1, 2, \ldots$ (or for a mathematician the numbers themselves) are put into a box as follows. At 1 minute to noon the numbers 1 to 10 are put in, and the number 1 is taken out. At $\frac{1}{2}$ minute to noon numbers 11 to 20 are put in and the number 2 is taken out. At $\frac{1}{3}$ minute to noon 21 to 30 in and 3 out; and so on. How many are in the box at noon? The answer is none: any selected number, e.g. 106, is absent, having been taken out at the 106th operation.

*An analyst is constantly meeting just such things; confronted with the set of points

$$P_1 + P_2 + \ldots + P_{10} - P_1 + P_{11} + \ldots + P_{20} - P_2 + \ldots,$$

he would at once observe that it was 'null', and without noticing anything paradoxical.*

On the subject of paradoxes I will digress into Celestial Mechanics. Suppose n bodies, to be treated as points, are moving subject to the

Newtonian law of gravitation. Those systems are infinitely rare[1] for which, sooner or later, a simple collision (collision of two point-bodies only) occurs. It is as certain as anything can be that the same holds for multiple collisions (of three or more point-bodies). (Indeed, while simple collisions are normal for e.g. the inverse cube law, multiple ones are doubtless infinitely rare whatever the law.) Nevertheless there is no proof.

This is of course a paradox about proofs, not about facts. It is also possible to explain it. With simple collisions the analytic character of the behaviour of the system, suitably generalized, survives a simple collision, and it can consequently be seen that a simple collision (at no matter how late a date) involves two analytic relations between the initial conditions, and this makes those conditions infinitely rare.

§2. (6) 4 ships A, B, C, D are sailing in a fog with constant and different speeds and constant and different courses. The 5 pairs A and B, B and C, C and A, B and D, C and D have each had near collisions; call them 'collisions'. Most people find unexpected the mathematical consequence that A and D necessarily 'collide'. Consider the 3-dimensional graphs of position against time — 'world lines' — with the axis of time vertical. The world lines a, b, c meet each other; consequently a, b, c are all in the same one plane, p say (see Fig. 1); d meets b and c, so it lies in p, and therefore meets a. (Multiple collisions are barred. Limiting cases of parallelism are ruled out by the speeds being different.)

(7) *An experiment to prove the rotation of the earth.* A glass tube in the form of an anchor ring is filled with water and rests horizontally, for simplicity at the North Pole. The tube is suddenly rotated through 180° about a horizontal axis. The water is now flowing round the tube (at the rate of a revolution in 12 hours) and the movement can be detected. This might have been invented by Archimedes, but had to wait till 1930 (A. H. Compton). (It is curious how very late many of the things in my collection are. For a change the date of the next one is 1605.)

(8) *Stevinus and gravity on an inclined plane.* A chain $ABCD$, hanging as in Fig. 2, *can* rest in equilibrium (else perpetual motion). The symmetrical lower part ADC exerts equal pulls on AB, BC, and may be removed, leaving ABC in equilibrium. That AB, BC balance is in effect the sine law. (For an interesting discussion see Mach, *Science of Mechanics*, 24–31.)

---

[1]In technical language the set of 'representative points' (in a space for representing initial conditions) of systems that suffer simple collisions has measure 0.

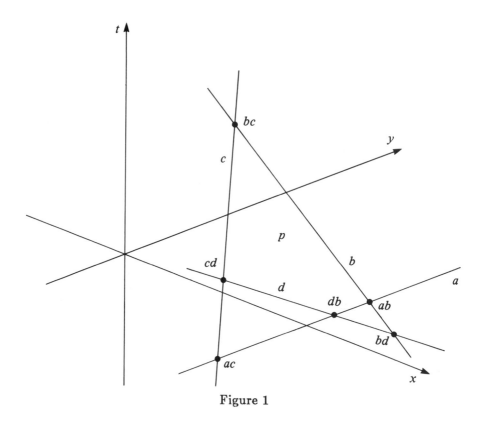

Figure 1

(9) To determine the orbit of a planet or comet 3 observations, each of two (angular) co-ordinates and the time t, suffice. It is actually the case that to *any* set of observations (point the telescope anyhow at any 3 times) an orbit[1] corresponds. Imagine a speck on the telescope's object glass; this satisfies the observations, and it also describes an orbit (that of the earth). Now (some sordid details being taken for granted) the equations for the elements of the orbit are reducible to the solution of an equation of the 8th degree. This has accordingly one real root. But since the degree is even it must have a second real root. This to all intents rigorous argument is a test of taste. Incidentally the joke is *in* the mathematics, not merely about it.

(10) *Dissection of squares and cubes into squares and cubes, finite in number and all unequal.* The square dissection is possible in an infinity of distinct ways (the simplest due to Duijvestijn, is shown in Fig. 3); a cube dissection is impossible. The surprising

---

[1] A conic with the sun as focus

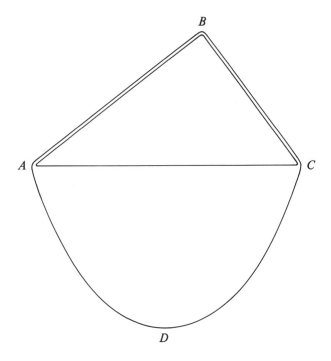

Figure 2

proof of the first result is technical. (See R. Sprague, 'Beispiel einer Zerlegung des Quadrats in lauter verschiedene Quadräte,' J. Reine Angew. Math. **182** (1940), 60–64 and R. L. Brooks, C. A. B. Smith, A. H. Stone and W. T. Tutte, 'The Dissection of Rectangles into Squares,' *Duke Mathematical Journal*, **7** (1940), 312–340.) Brooks et al give the following elegant proof of the second. In a square dissection the smallest square is not at an edge (for obvious reasons). Suppose now a cube dissection does exist. The cubes standing on the bottom face induce a square dissection of that face, and the smallest of the cubes at the face stands on an internal square. The top face of this cube is enclosed by walls; cubes must stand on this top face; take the smallest — the process continues indefinitely. (If, however, we ask whether a single cube can be completely surrounded by larger unequal ones, the answer is 'yes'.)

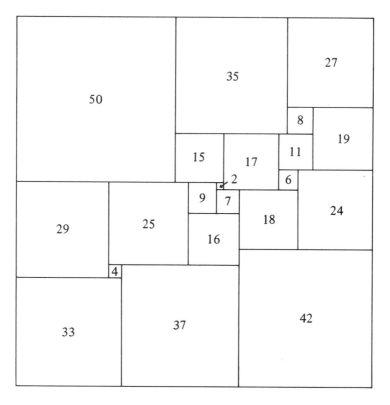

Figure 3

(11) *A voting paradox.* If a man abstains from voting in a General Election on the ground that the chance of his vote mattering is negligible, it is common to rebuke him by saying 'Suppose everyone acted so.' The unpleasant truth that the rebuke is fallacious in principle is perhaps fortunately hidden from the majority of the human race. Consider, however, the magnitudes involved, where the election and the constituency are reasonably open. The chance that his vote will elect his member by a majority of 1 is of the order of 1 in 5000; there is a further chance of the order of 1 in 50 that this result will cause a change of Government. The total chance for this is no worse than 1 in 250,000. Since there are 30,000,000 voters with similar opportunities it would appear that there is something wrong, the explanation is that when the event happens to one man, 20,000 or so[1] other voters in his constituency are in the same

[1] Half 70 per cent of 60,000

position.

A suggestion made in 1909, that two parties in the proportion p to q will have representations as p^3 to q^3, was revived by the *Economist* in January 1950; various earlier elections fit the rule very well. That the proportion is likely to be magnified is obvious from the fact that if the parties were peppered at random over the country a minority of 49 to 51 would not obtain a single seat. So it is a question of a pattern of localized interests, not of a single mysterious cause.

I am prepared to debunk the rule. In the first place it may be replaced by the simpler one that the percentages $50 \pm x$ of the parties should have their difference magnified by a constant c to give representations $50 \pm cx$. With $c = 3$ the two rules agree substantially up to $x = 6$ and then begin to diverge. (With 600 seats a majority 56: 44 gets 405 seats under the cube rule, 408 under the 3-rule; with 57 : 43 the figures become 420 and 426.) This probably covers all practical cases. (For wider differences one would expect extinction of the minority at some point; the 3-rule predicts it at 67 : 33.) Next, it is not unreasonable that with a given pattern of constituencies there should be a magnification from $50 \pm x$ to $50 \pm cx$ in the limit as x tends to 0, and this might be extended to work over a range like 0 to 6 of x by the familiar process of cooking c to fit the far end of the range. A change in the constituency-pattern might alter c (the above was written before the election of February 1950, which had a new pattern and a new c). The roundness of the number 3 probably impresses; but unduly; it might almost as easily be 2.9; and after all the velocity of light and the gravitation constant start off with 300 and 666.

(12) *The problems on weighing pennies.* This attractive and 'genuinely mathematical' subject was exhaustively discussed a few years ago, and I will do no more than mention it. (See in particular the masterly analysis by C. A. B. Smith, *Mathematical Gazette*, **XXXI** (1947), 31–39.)

§3. As I explained in §1 there are things, more 'mathematical' than most of the foregoing, with high claims but omitted on the ground of 'familiarity'. From topology we may mention (13) the Möbius strip, (14) the Klein bottle, (15) the four colour problem, and (16) the Jordan curve problem (the last two are extreme, very extreme, instances where the result is easy to understand while the proof is very difficult). The 'fixed point theorems' belong here (and are discussed in §6). There is an attractive chapter on topology in *What is Mathematics* by Courant and Robbins (*CR* for reference), to which the interested reader may be referred.

I mention also, and again without discussion, two pioneering discoveries of Cantor, (17) the non-denumerability of the continuum, and (18) the fact that the points of a line and those of a square are 'similar classes', i.e. can be 'paired off' — put into 'one-one correspondence' (*CR*, 81–85).

(19) The (Schröder-Bernstein) theorem: if a class A is similar to a sub-class A_1 of itself, then it is similar to any ('intermediate') sub-class containing A_1. In this case not only does the result use no more raw material than classes, but there is a proof which brings in no further ideas (except common sense ones like 'all'); it does not mention numbers, or sequences, and is unaware of the concept 'finite' (and its negation 'infinite'). None the less the proof imposes on this simplest of raw material a technique which is too much for the amateur.

At this point there belong the 'reflexive paradoxes'. For Russell's original contradiction about classes, see *CR*, 87, and for two more see p. 58 below (where they figure under the heading of 'jokes'!).

We now begin to part company more frequently from the amateur, and I will stop numbering.

* §4. *An isoperimetrical problem*: an area of (greatest) diameter not greater than 1 is at most $\frac{1}{4}\pi$.

Proofs of various lengths exist. It is easy to see that we may suppose the area convex, and on one side of one of its 'tangents'. With polar coordinates as in Figure 4,

$$\text{area} = \tfrac{1}{2} \int_0^{\frac{1}{2}\pi} \left(r^2(\theta) + r^2(\theta - \tfrac{1}{2}\pi) \right) d\theta.$$

The integrand is $OP^2 + OQ^2 = PQ^2 \le 1$.

Suppose $a_n > 0$ for all n. Then

$$\varlimsup_{n \to \infty} \left(\frac{1 + a_{n+1}}{a_n} \right)^n \ge e,$$

and the result is best possible. (From Pólya)*

A rod is hinged to the floor of a railway carriage and let go at random; there is then a small but nonzero probability that, uninterfered with, it will still be standing up at the end of a fortnight: the chance is about 1 in 10^{10^5}. (The train is not an 'ideal' one: it starts e.g. from King's Cross at 3.15, proceeds to a tunnel where it stops for 5 minutes; after a dozen or so further stops it reaches Cambridge at 5.35. I seem

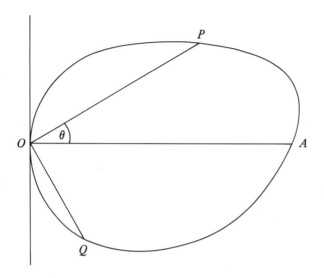

Figure 4

to remember being told that the genius who asked the original question was unable to answer it.)

There is a proof in *CR*, 319. *Alternatively we may argue (with reasonable latitude of interpretation) as follows. Consider an initial position with the rod at relative rest and making an angle θ measured from the *left* horizontal (to avoid later clashes of meanings of left and right). Let S be the set of initial positions θ for which the rod sooner or later dies down on the left. Subject to very slight assumptions about the circumstances of the journey and the interaction of train and rod we have the governing fact that *the set S is open* (and we need not try to lay down detailed assumptions). Let θ_0 be the least upper bound of the angles θ of S. Then θ_0 is not a member of S and from initial θ_0 the rod does not lie down on the left. If on the other hand it lies down on the right it will do so also for all θ near enough to θ_0 on the left; this is false since some (actually all) of these θ belong to S. So from θ_0 the rod never lies down. And for some small enough sector about this θ_0 it will not lie down within a fortnight.

It is instructive to consider why the argument does not similarly prove that the rod, properly started, never moves more than say half a degree from the initial position.*

This argument is instantly convincing to the mathematician. Of possible variants the one I choose seems best suited to be interpreted to the amateur.

Suppose we have any collection S (in general infinite in number) of numbers, or points of a line, 'bounded on the right'; that is to say there are points P (far enough to the right) such that no member of S is to the right of P (the statement deliberately contemplates the possibility of a member *coinciding* with P). Any such P is called an upper bound (u.b. for short) of the set S. If the event happens for P it happens (a fortiori) if P is moved to the right. If the event happens for P_0, *but not if P_0 is moved, however little this movement, to the left*, P_0 is called the *least* upper bound (l.u.b.) of S. (E.g. 1 is the l.u.b. of the set of numbers, or again of the set of rational numbers, between 0 and 1, with 1 excluded; and, again, 1 is the l.u.b. of these two sets modified by 1 being *included*. In the first pair of cases the l.u.b. is not a member of S, in the second pair it is.) After a little mental experiment it should be intuitive that *every set S bounded on the right has a l.u.b.* Note that there are two defining properties of a l.u.b.: (1) it is a u.b., (2) anything to the left is not.

So far the rod-problem has not entered (and the successful reader has acquired an important mathematical conception and theorem). In the argument that follows the various steps can be usefully checked against the special case of the train at rest; there we know the answer, the rod, started vertical, never moves.

Suppose the rod is started at relative rest at an angle θ measured from the *left* horizontal; call this 'initial position θ'. Consider the set, call it S, of initial positions θ from which the rod lies down sooner or later on the left. If it does this for a particular Q, then *it will do so also for all near enough to it on either side*; any θ belonging to S is in the middle of a block of θ all belonging to S (in mathematical language 'S is open'). This intuitive fact, on which everything turns, depends on very slight assumptions about the various circumstances, and we need not try to state them in detail.

Let θ_0 be the l.u.b. of the set S. Then θ_0 is not a member of S, for if it were, near enough θ on the right would belong to S and θ_0 would not be an u.b. of S. So the rod, started at θ_0 does not lie down to the left. If on the other hand it lies down to the right, it would do so also from *all* near enough θ on the *left* (the 'open' principle), so that none of these can be members of S; this, however, means that θ_0 could be moved to the left without ceasing to be an u.b. of S, contrary to property (2) of a l.u.b. So from Q_0 the rod never lies down. And from some small enough sector round θ_0 it does not lie down within a fortnight.

A similar result is true if the hinge of the rod is replaced by a universal joint. This time, however, we have to appeal to a high-brow 'fixed point theorem' (*CR*, 321).

A typical fixed point theorem is the following. A thin rubber sheet covering the surface of a sphere is deformed without tearing or overlapping. Suppose now further that no point P has its new position P' diametrically opposite its old one. The theorem now is that there must be at least one 'fixed point' (i.e. a P for which P', the 'transform' of P, is the same as P).

The amateur will probably agree that this is elegant (the mathematician says 'beautiful'). But, what he can hardly be expected to realize, it is not only important in itself, but has profound consequences in unexpected fields (like Celestial Mechanics). Importance *plus* simplicity (i.e. simplicity of *result*) give the fixed point theorems very high claims indeed. The status of the *proofs*, however, is something we have not as yet come across. In the first place they took a great deal of finding; indeed Poincaré, while originating the 'fixed point' conception, stating some of the theorems, and fully aware of their prospective consequences, did not himself *prove* anything. On the other hand the proofs, *once found*, are not beyond the amateur. (There is one in *CR*, see pp. 251–255.)

Is the theorem for the sphere intuitive? I think it can be made so. Suppose there is no fixed point (most of the proofs start like this). Then there is a (unique) piece of great circle PP' from each P of the sphere to its transform P', and its direction (from P) varies continuously with P. (Note incidentally that we are here *using* the assumption that P' is not diametrically opposite P; if it were, the arc PP' would not be unique.) Suppose now that the sphere is covered with hair. If the hair with its root at P is made to fall along PP' the sphere has succeeded in brushing its hair smoothly, with no 'singular' points of 'parting' or 'meeting': we know intuitively that this is impossible. So there must be at least 1 singular point, itself not provided with a hair. Contrary to a hasty impression there need not be *more* than 1, this serving as both parting and meeting. Fig. 5 shows how the hair directions run. A dog, which is roughly a topological sphere, makes no attempt to economize in this way; it has a line of parting down its back and one of 'meeting' below.

Every convex closed analytic surface must possess at least two umbilics[1] ($R_1 = R_2 = \infty$ is permitted, and it is possible for there to

---

[1] An umbilic is a point near which the surface is approximately spherical (or plane); the two 'radii of curvature' are equal.

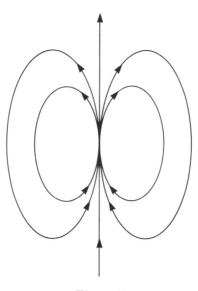

Figure 5

be only two, and of this kind). The remarkable feature of this theorem is that the only existing proof occupies 180 pages.

[This is no longer quite true, though the proofs are still very long. It is unlikely that anybody has really read H. L. Hamburger's papers: *Ann. Math.* **41** (1940), 63–86 and *Acta Math.* **73** (1941), 174–332. A simpler proof was given by G. Bol, Math. Z. **49** (1943/44), 389–410, which was not complete but was improved by T. Klotz, *Comm. Pure and Applied Math.* **12** (1959), 277–311. Another proof is due to C. J. Titus, *Acta Math.* **131** (1973), 43–77.]

§5. What is the best stroke ever made in a game of Billiards? Non-mathematical as this sounds, I claim that, granted the question can be asked significantly, the nature of the answer is deducible by reasoning. [1] It might indeed be doubted whether the stroke is *possible*, but it did happen that Lindrum, having in the middle of a long break left the object white over a pocket, deliberately played to make a cannon in which the white balls were left touching, and succeeded. (The balls were spotted in accordance with the laws, and the break could continue.)

A flexible inextensible roll (e.g. of film) has its free edge attached horizontally to an inclined plane and uncoils under gravity (Fig. 6). The

---

[1] I should be glad to think that when a reader approves of an item he is agreeing with me in finding the 'point', one specially congenial to a mathematician.

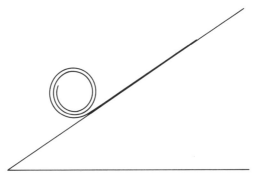

Figure 6

line of contact has always zero velocity, and no kinetic energy is destroyed during the motion: when the roll has completely uncoiled there has been a loss of potential energy and the kinetic energy is zero; what has happened to the missing energy?

There is an analogy in daily life; the crack of a whip. With an 'ideal' whip the motion of the tip ends with a finite tension acting on zero mass and the velocity becomes infinite. In practice the velocity does exceed the velocity of sound and a crack results. Perhaps the nearest approach to infinity in everyday life.

A weight is attached to a point of a rough weightless hoop, which then rolls in a vertical plane, starting near the position of unstable equilibrium. What happens, and is it intuitive?

The hoop lifts off the ground when the radius vector to the weight becomes horizontal. I don't find the lift directly intuitive; one can, however, 'see' that the motion is equivalent to the weight's sliding smoothly under gravity on the cycloid it describes, and it is intuitive that it will sooner or later leave *that*. (But the 'seeing' involves the observation that W is instantaneously rotating about P (Fig. 7).)

Mr H. A. Webb sets the question annually to his engineering pupils, but I don't find it in books.

In actual practice the hoop skids first.

Suppose buses on a given route average 10-minute intervals. If they run at exactly 10-minute intervals the average time a passenger arriving randomly at a stop will have to wait is 5 minutes. If the buses are irregular the average time is greater; for one kind of random distribution it is 10 minutes, and, what is more, the average time since the previous bus is also 10 minutes. For a certain other random distribution, both times become *infinity* (the long waits dominate in spite of their rarity).

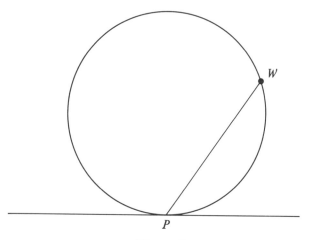

Figure 7

The two things I mention next are alike in that a first guess at the odds is almost certain to be wrong (and they offer opportunities to the unscrupulous). The first has a number of forms (one of them — letters in wrong envelopes — is the source of the sub-factorial notation); the latest is as follows. From two shuffled packs the two top cards are turned up, then the next two, and so on. One bets that sooner or later the pair of cards will be the 'same' (e.g. both 7's of spades). This is fairly well known, but most people who do not know it will offer good odds against; actually the odds are approximately $17 : 10$ on (practically $e - 1 : 1$).

In the other we have a group of 23 people; what are the odds that some pair of them have the same birthday?[1] Most people will say the event is unlikely, actually the odds are about evens.

Kakeya's problem. Find the region of least area in which a segment of unit length can turn continuously through 360° (minimize the area swept over). It was long taken for granted that the answer was as in Fig. 8 and the area $\frac{1}{8}\pi$. In 1930, however, A. S. Besicovitch *(Math. Zeit.* **27** (1928), 312–320) showed that the answer is zero area (unattained): given an arbitrarily small ϵ the area swept can be less than ϵ. As ϵ tends to 0 the movements of the segment become infinitely complicated and involve excursions towards infinity in all directions.

Crum's problem. What is the maximum number of convex polyhedra of which every pair have a face (or part of one) in com-

---

[1] In the usual sense: they may have different ages.

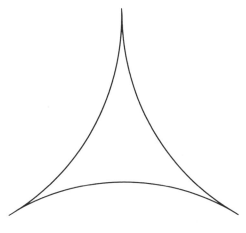

Figure 8

mon? In the corresponding problem in 2 dimensions the answer is fairly easily seen to be 4 (see Fig. 9); the natural expectation in 3

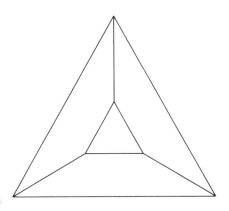

Figure 9

dimensions is 10 to 12. The answer was given in 1905 by Tietze and rediscovered by Besicovitch in 1947 (*J.L.M.S.* **22**, 285–287). In Besicovitch's case it is a foil to his previous problem; there he annihilated something; here he does the opposite — the answer is infinity.

The question recently arose in conversation whether a dissertation of 2 lines could deserve and get a Fellowship. I had answered this for myself long before; in mathematics the answer is yes.

*Cayley's projective definition of length is a clear case if we may interpret '2 lines' with reasonable latitude. With Picard's Theorem it could be literally 2, one of statement, one of proof.

(Theorem.) An integral function never 0 or 1 is a constant.

(Proof.) $\exp\{i\Omega(f(z))\}$ is a bounded integral function.

$(\tau = \Omega(w)$ inverse to $w = k^2(\tau))$.

The last bracket is needed solely because of the trivial accident that the function Ω, unlike its inverse $k^2(\tau)$, happens to have no unmistakable name.

The function $k^2(\tau)$ is the *modular* function which arises from the theory of elliptic functions. It gives an analytic map from the half-plane $\{\tau \in \mathbf{C} : \operatorname{Im}\tau > 0\}$ onto $\mathbf{C}\setminus\{0,1\}$. The inverse Ω is many-valued but, for any branch of it, $\Omega(f(z))$ extends analytically to give an integral function from $\mathbf{C}$ into $\{\tau \in \mathbf{C} : \operatorname{Im}\tau > 0\}$.*

With Cayley the importance of the idea is obvious at first sight. With Picard the situation is clear enough today (innumerable papers have resulted from it). But I can imagine a referee's report: 'Exceedingly striking and a most original idea. But, brilliant as it undoubtedly is, it seems more odd than important; an isolated result, unrelated to anything else, and not likely to lead anywhere.'

*Euclid's proof that there is an infinity of primes can be condensed, for the professional, into one line. 'If $p_1, p_2, \ldots, p_n$, $1 + p_1 p_2 \ldots p_n$ is not divisible by any p_m.'

So can the proof of H. Bohr's famous result that '$\varsigma(s)$ is unbounded in $\sigma > 1$ for large t.'

$$\varlimsup_{\sigma>1, t\to\infty} \operatorname{Re}\varsigma(\sigma + it) \geq \varlimsup_{\sigma\to 1+0}\ \varlimsup_{t\to\infty} \sum n^{-\sigma} \cos(n \log t)$$

$$= \varlimsup_{\sigma} \sum n^{-\sigma} = \infty.$$

My last example is a high-brow theorem in analysis.[1] But any mathematician willing to take the importance of the result for granted should be able by judicious skipping to follow the essentials of the proof:[2] this turns on an idea (see the passages in italics) the most impudent in mathematics, and brilliantly successful.

---

[1] The penultimate step forces the reader to see that 'Dirichlet's theorem' is being used, and to make the necessary extension of it.

[2] This begins at p. 41, line −2, and he can take the lemma for granted.

A very important theorem (due to M. Riesz) is as follows.

For $\alpha, \beta > 0$ let $M_{\alpha\beta}$ be the least upper bound (l.u.b. for short) of

$$|L(x, y)| = \left| \sum_{\mu=1}^{m} \sum_{\nu=1}^{n} a_{\mu\nu} x_{\mu} y_{\nu} \right|$$

for constant complex $a_{\mu\nu}$ and varying complex x_{μ}, y_{ν} subject to

$$\sum |x_{\mu}|^{1/\alpha} \leq 1, \qquad \sum |y_{\nu}|^{1/\beta} \leq 1.$$

Then $\log M_{\alpha\beta}$ is a convex function of α, β (in $\alpha, \beta > 0$).

A convex function of one variable has a graph in which (in the wide sense) 'the arc is below the chord'. With several variables $\alpha, \beta, \ldots$ the function is to be convex in σ on any line l, $\alpha = \alpha_0 + \lambda_1 \sigma, \beta = \beta_0 + \lambda_2 \sigma, \ldots$ For the purpose of applications the theorem is extended to a form, T for short, which has a different outward appearance and takes a lot of proving, but the above is the essential foundation. In T the ranges of α and β are extended to include 0, and it then enables us to 'interpolate' in a very drastic manner between a pair of known theorems. Thus, the 'Young-Hausdorff' inequality

$$\left(\sum |c_n|^{p/(p-1)} \right)^{(p-1)/p} \leq \left(\frac{1}{2\pi} \int_{-\pi}^{\pi} |f|^p \, d\theta \right)^{1/p}$$

for the Fourier constants c_n, of a function $f(\theta)$ of L^p is in fact valid for $1 \leq p \leq 2$ (with a crude 'interpretation' for $p = 1^1$). T enables us to assert the general inequality if only we know it for $p = 1$ and $p = 2$.[2] For $p = 2$ it reduces to Bessel's inequality, and at $p = 1$ (such is the power of T) we need to know only the crude form, and this is trivial. T thus produces a high-brow result 'out of nothing'; we experience something like the intoxication of the early days of the method of projecting conics into circles.

Until lately there was no proof of the $M_{\alpha\beta}$ theorem that was not very decidedly difficult. The one I now present is due to G. Thorin ('Convexity Theorems': *Communications du séminaire mathématique de l'Université de Lund*, 9).

We use three immediately obvious principles; (a) the l.u.b. of a family (possibly infinite in number) of convex functions is convex; (b)

---

[1] Namely 'l.u.b. $|c_n| \leq \frac{1}{2\pi} \int |f| \, d\theta$'.
[2] The l concerned joins the points $(\frac{1}{2}, \frac{1}{2})$, $(1, 0)$, α is $1/p$, β is $(p-1)/p$.

a limit of a sequence of convex functions is convex; (c) in finding the l.u.b. of something subject to a number of independent conditions on the variables we may take the conditions in any order (or simultaneously). E.g. quite generally

$$\underset{0\le x,y\le 1}{\text{l.u.b.}} |f(x,y)| = \underset{0\le x\le 1}{\text{l.u.b.}} \left(\underset{0\le y\le 1}{\text{l.u.b.}} |f(x,y)| \right)$$

It follows from (a) and (c) in combination that if we can express $\log M_{\alpha\beta}$ as

$$\text{l.u.b.}(\text{l.u.b.}(\ldots(\text{l.u.b.}|L(x,y)|)\ldots)),$$

in such a way that the *innermost* l.u.b. (with all the variables for the outer ones fixed) is convex in (α, β), we shall have proved the theorem. Thorin, however, *takes his (innermost) l.u.b. with respect to a variable that is not there!*

We must begin with:

Lemma. Suppose that $b_1, b_2, \ldots, b_N$ are real and that $f(s)$ is a finite sum $\sum a_r e^{b_r s}$ (or more generally an integral function of $e^{b_1 s}, \ldots, e^{b_N s}$), where $s = \sigma + it$. Let $m(\sigma) = \text{l.u.b.}|f(s)|$. Then $\log m(\sigma)$ is convex in σ.

By principle (b) it is enough to prove this when the b's are rationals β'/β. Then if D is the L.C.M. of the β's, f is an integral function of $ze^{s/D}$. Hadamard's 'three circles theorem', that '$\log M(r)$ is convex in $\log r$', now gives what we want.

Come now to the theorem. We have to prove $\log M_{\alpha\beta}$ convex on every interval l, or $\alpha = \alpha_0 + \lambda_1\sigma$, $\beta = \beta_0 + \lambda_2\sigma$, contained in $\alpha, \beta > 0$. For such α, β we may write

$$x_\mu = \xi_\mu^\alpha e^{i\phi_\mu}, \quad y_\nu = \eta_\nu^\beta e^{i\psi_\nu}, \quad \xi, \eta \ge 0;$$

and then, for varying (real) ϕ, ψ and (real) ξ, η varying subject to (1)[1]

$$\sum \xi \le 1, \quad \sum \eta \le 1, \quad \xi\eta \ge 0,$$

we have

$$M_{\alpha\beta} = \underset{(\phi,\psi,\xi,\eta)}{\text{l.u.b.}} \left(\left| \sum\sum \alpha_{\mu\nu} \xi_\mu^{\alpha_0 + \lambda_1\sigma} \eta_\nu^{\beta_0 + \lambda_2\sigma} e^{i(\phi_\mu + \psi_\nu)} \right| \right).$$

If in this we replace σ by $s = \sigma + it$ (for any real t) the l.u.b. is unaltered (the maximal $\phi + \psi$'s being merely 'translated'). *We can now add an*

---

[1]These conditions are independent of α, β.

operation 'l.u.b. with respect to t'. By principle (c) we make this the innermost one: summing up and taking logarithms we have

$$\log M_{\alpha\beta} = \underset{(\phi,\psi,\xi,\eta)}{\text{l.u.b.}} \; \log m(\sigma; \phi, \psi, \xi, \eta),$$

where

$$m(\sigma) = \underset{(t)}{\text{l.u.b.}} \big|f(s)\big|,$$

$$f(s) = f(s; \phi, \psi, \xi, \eta) = \sum\sum \alpha \xi^{\alpha_0 + \lambda_1 s} \eta^{\beta_0 + \lambda_2 s} e^{i(\phi + \psi)}.$$

For the range l of α, β (in which the indices are positive), and fixed ϕ, ψ, ξ, η, we can suppress in the sum f any terms in which a ξ or an η is 0; *the modified f has the form of the f in the lemma, $\log m(\sigma)$ is convex for all σ, and in particular for the range concerned.* $\log M_{\alpha\beta}$ *is now convex in σ by principle (a).**

§6. *Ciphers.* The legend that every cipher[1] is breakable is of course absurd, though still widespread among people who should know better. I give a sufficient example, without troubling about its precise degree of practicability. Suppose we have a 5-figure number N. Starting at a place N in a 7-figure log-table take a succession of pairs of digits $d_1 d_1'$, $d_2 d_2' \ldots$ from the last figures of the entries. Take the remainder of the 2-figure number $d_n d_n'$ after division by 26. This gives a 'shift' s_n, and the code is to shift[2] the successive letters of the message by $s_1, s_2, \ldots$ respectively.

It is sufficiently obvious that a *single* message cannot be unscrambled, and this even if all were known except the key number N (indeed the triply random character of s_n is needlessly elaborate). If the same code is used for a number of messages it could be broken, but all we need do is to vary N. It can be made to depend on a date, given in clear; the key might e.g. be that N is the first 5 figures of the 'tangent' of the date (read as degrees, minutes, seconds: 28° 12′ 52″ for Dec. 28, 1952). This rule could be carried in the head, with nothing on paper to be stolen or betrayed. If any one thinks there is a possibility of the entire scheme being guessed he could modify 26 to 21 and use a date one week earlier than the one given in clear.

---

[1] I am using the word cipher as the plain man understands it.
[2] A shift of $s = 2$ turns 'k' into 'm', 'z' into 'b'.

FROM THE MATHEMATICAL TRIPOS

§1. It is always pleasant to find others doing the silly things one does oneself. The following appears as a complete question in Schedule B for 1924 (Paper I). (a)[1] An ellipsoid surrounded by frictionless homogeneous liquid begins to move in any direction with velocity V. Show that if the outer boundary of the liquid is a fixed confocal ellipsoid, the momentum set up in the liquid is $-MV$, where M is the mass of the liquid displaced by the ellipsoid.

(The result was later extended to other pairs of surfaces, e.g. two coaxial surfaces of revolution.)

Whatever the two surfaces are we can imagine the inner one to be filled with the same liquid; then the centre of mass does not move.

Published sets of examination questions contain (for good reasons) not what was set but what ought to have been set; a year with no correction is rare. One year a question was so impossibly wrong that we substituted a harmless dummy.

There used to be 'starred' questions in Part II (present style), easy, and not counting towards a 1st. A proposed starred question was once rejected, proposed and rejected as too hard for an unstarred one, and finally used as a question in Part III.

Once when invigilating I noticed, first that the logarithm tables provided did not give values either for e or for $\log e = 0.4343$, and secondly that question 1 asked for a proof that something had the numerical value 4.343 (being in fact $10 \log e$). Was I to announce the missing information, thereby giving a lead? After hesitation I did so, and by oversight committed the injustice of not transmitting the information to the women candidates, who sat elsewhere.

§2. I inherited Rouse Ball's 'Examiner's books' for the Triposes

---

[1] Let the amateur read bravely on.

round about 1881. Some details may be of interest. The examination, of 18 three-hour papers, was taken in January of the 4th year. In one year full marks were $33,541$, the Senior Wrangler got $16,368$, the second $13,188$, the last Wrangler 3123, the Wooden Spoon (number ninety odd) 247. The first question carried $6 + 15$ marks, a question of the 2nd problem paper 500 (more than twice the bottom score).

As a staunch opponent of the old Tripos I was slightly disconcerted to find a strong vein of respectability running throughout. It is surprising to discover that a man who did all the bookwork (which was much the same as it is now) and nothing else would have been about 23rd Wrangler out of 30. Since even the examiners of the '80s sometimes yielded to the temptation to set a straightforward application of the bookwork as a rider, he would pick up some extra marks, which we may suppose would balance occasional lapses. The two heavily marked problem papers contained of course no bookwork for him to do; if we suppose that he scored a quarter of the marks of the Senior for these papers, or say 7 per cent of the total, he would go up to about 20th. (Round about 1905 the figures would be 14th Wrangler out of 26 for pure bookwork, rising to 11th on 7 per cent. of the problem papers, and incidentally straddling J. M. Keynes.)

§3. On looking through the questions, and especially the problem papers, for high virtuosity (preferably vicious) I was again rather disappointed; two questions, however, have stuck in my memory.

(b) A sphere spinning in equilibrium on top of a rough horizontal cylinder is slightly disturbed; prove that the track of the point of contact is initially a helix.

*Pursuing this idea an examiner in the following year produced (Jan. 18, morning, 1881, my wording).

(c) If the sphere has a centrally symmetrical law of density such as to make the radius of gyration a certain fraction of the radius then, whatever the spin, the track is a helix so long as contact lasts. (Marked at 200; a second part about further details carried another 105.)*

The question about (b) is whether it can, like (a), be debunked. On a walk shortly after coming across (b) and (c) I sat down on a tree trunk near Madingley for a rest. Some process of association called up question (b), and the following train of thought flashed through my mind. 'Initially a helix' means that the curvature and the torsion are stationary at the highest point P; continue the track backwards; there is skew-symmetry about P, hence the curvature and torsion are stationary.

I now ask: is this a proof, or the basis of one, how many marks should I get, and how long do you take to decide the point?

*§4. Proceed to (c). I do not regard this question[1] as vicious: it involves the general principles of moving axes with geometrical conditions; a queer coincidence makes the final equations soluble, but this is easy to spot with the result given. The extremely elegant result seems little known.

Take moving axes at the centre of the sphere with Oy along the normal to the point of contact, making an angle, θ say, with the vertical, and Oz parallel to the cylinder. Eliminating the reactions at the point of contact we get (cf. Lamb, *Higher Mechanics*, 165–166):

$$(I + Ma^2)b\ddot{\theta} = Ma^2 g \sin\theta,$$
$$(I + Ma^2)\dot{\omega} = Ia\dot{\theta}q,$$
$$a\dot{q} + \omega\dot{\theta} = 0,$$

which, on normalizing to $a = 1$, become respectively, say,

$$\lambda^2\ddot{\theta} = \sin\theta, \quad \mu^2\dot{\omega} = \dot{\theta}q, \quad \dot{q} = -\omega\dot{\theta}.$$

The 2nd and 3rd of these lead to,

$$q = n\cos(\theta/\mu), \quad \mu\omega = n\sin(\theta/\mu),$$

where n is the initial spin. If $\mu = 2$ these combine with the first to give $\omega = \frac{1}{4}n\lambda\dot{\theta}$ and so $z = \frac{1}{4}n\lambda\theta$.* Suppose a sphere is started rolling on the inside of a rough *vertical* cylinder (gravity acting, but no dissipative forces); what happens? The only sensible guess is a spiral descent of increasing steepness; actually the sphere moves up and down between two fixed horizontal planes. Golfers are not so unlucky as they think.

Some time about 1911 an examiner A proposed the question: E and W are partners at Bridge; suppose E, with no ace, is given the information that W holds an ace, what is the probability p that he holds 2 at least? A colleague B, checking A's result, got a different answer, q. On analysis it appeared that B calculated the probability, q, that W has 2 aces at least given that he has the spade ace. p and q are not the same, and $q > p$.

---

[1] The actual question gave the law of density and left the radius of gyration to be calculated, and asked for some further details of the motion.

Subject always to E's holding no ace, $1 - q$ is the probability of W holding S ace only, divided by the probability of his holding at least S ace; $1 - p$ is the probability of his holding 1 ace only, divided by the probability of his holding at least 1 ace. The 2nd numerator is 4 times the 1st. Hence

$$\frac{1-p}{1-q} = \frac{(\text{prob. of } S \text{ ace at least}) + \ldots + (\text{prob. of } C \text{ ace at least})}{(\text{prob. of } S \text{ ace at least})}$$

Since the contingencies in the numerator overlap, this ratio is greater than 1.

The fallacy '$p = q$' arises by the argument: 'W has an ace; we may suppose it is the spade'. But there is no such thing as 'it'; if W has more than one the informer has to *choose* one to be 'it'. The situation becomes clearer when a hand of 2 cards is dealt from the 3 cards, S ace, H ace, C 2. Here we know in any case that the hand has an ace, and the probability of 2 aces is $\frac{1}{3}$. If we know that the hand has the S ace then the probability of 2 aces is $\frac{1}{2}$.

CROSS-PURPOSES, UNCONSCIOUS ASSUMPTIONS, HOWLERS, MISPRINTS, ETC.

§1. A good, though non-mathematical, example is the child writing with its left hand 'Because God the Father does.' (He has to; the Son is sitting on the other one.)

I once objected to an apparently obscure use of the phrase 'Let us suppose for simplicity.' It should mean that the writer could do the unsimplified thing, but wishes to let the reader off; it turned out that my pupil meant that he had to simplify before *he* could do it.

It is of course almost impossible to guard against unconscious assumptions. I remember reading the description of the coordinate axes in Lamb's *Higher Mechanics*: Ox and Oy as in 2 dimensions, Oz vertical. For me this is quite wrong; Oz is horizontal (I work always in an armchair with my feet up).

How would the reader present the picture of a closed curve (e.g. a circle) lying entirely to one side of one of its tangents? There are 4 schools; I belong to that of a vertical tangent with the curve on the right; I once referred to the configuration, and without a figure, in terms that made nonsense for the other 3 schools.

How not to

A brilliant but slapdash mathematician once enunciated a theorem in 2 parts, adding: 'Part 2, which is trivial, is due to Hardy and Littlewood.'

The trivial part 2 needed to be stated 'for completeness', and Hardy and Littlewood had similarly needed to state it. The author had then to comply with the rule that nothing previously published may be stated without due acknowledgement.

In presenting a mathematical argument the great thing is to give the educated reader the chance to catch on at once to the momentary point and take details for granted: his successive mouthfuls should be such as can be swallowed at sight; in case of accidents, or in case he wishes for once to check in detail, he should have only a clearly circumscribed little problem to solve (e.g. to check an identity: *two* trivialities omitted can add up to an *impasse*). The unpractised writer, even after the dawn of a conscience, gives him no such chance; before he can spot the point he has to tease his way through a maze of symbols of which not the tiniest suffix can be skipped. I give below an example (from analysis, where the most serious trouble occurs). This is not at all extreme for a draft before it has been revised by some unfortunate supervisor or editor. It is unduly favourable to the criminal since the main point is hard to smother. But it is not easy to be interestingly boring, and in momentary default of a specimen of the genuine article it is the best I can do. (There is a game of selecting a team of the 11 most conspicuous representatives of a given quality: who are the 11 most brilliantly dim persons? My team is too blasphemous, seditious and libellous to quote.)

*A famous theorem of Weierstrass says that a function $f(x_1, x_2)$ continuous in a rectangle R, can be uniformly approximated to by a sequence of polynomials in x_1, x_2. It is valid in n dimensions, and the beginner will give what follows, but in $x_1, x_2, \ldots, x_n$; $x'_1, x'_2, \ldots, x'_n$. The proof, an audacious combination of ideas, is in two parts; the second cannot be badly mauled and I give it at the end. Here is the beginner's proof of the first part. I am indebted to Dr Flett for one or two happy misimprovements, and for additional realism have left some incidental misprints uncorrected.

With $f(x_1, x_2)$ continuous in $(-a \le x_1 \le a, -b \le x_2 \le b)$, let $c > 0$ and define a function $f_1(x_1, x_2)$ by

$$f_1(x_1, x_2) = \begin{cases} f(-a, b) & (-c - a \le x_1 \le -a, b \le x_2 \le b + c) \\ f(x_1, b) & (-a \le x_1 \le a, b \le x_2 \le b + c) \\ f(a, b) & (-a \le x_1 \le a + c, b \le x_2 \le b + c) \\ f(-a, x_2) & (-a - c \le x_1 \le -a, -b \le x_2 \le b) \\ f(x_1, x_2) & (-a \le x_1 \le a, -b \le x_2 \le b) \\ f(a, x_2) & (a \le x_1 \le a + c, -b \le x_2 \le b) \\ f(a, -b) & (-a - c \le x_1 \le -a, -b - c \le x_2 \le -b) \\ f(x_1, -b) & (-a \le x_1 \le a, -b - c \le x_2 \le -b) \\ f(-a, -b) & (-a - c \le x_1 \le -a, -b - c \le x_2 \le -b) \end{cases}$$

It can easily be shown that $f_1(x_1, x_2)$ is continuous in

$$(-a - c \le x_1 \le a + c, -b - c \le x_2 \le b + c).$$

For points (x_1, x_2) of R define

$$\phi_n(x_1, x_2) =$$
$$\pi^{-1} n \int_{-a-c}^{a+c} dx_1' \int_{-b-c}^{b+c} f_1(x_1', y_1') \exp \left[-n\{(x_1' - x_1)^2 + (y_1' - y_1)^2\} \right] dx_2'.$$

We shall show (this is the first half referred to above) that

(1) $\phi_n(x_1, x_2) \to f(x_1, x_2)$ as $n \to \infty$, uniformly for (x_1, x_2) of R.

There is a $\delta(\epsilon)$ such that $|f_1(x_1'', x_2'') - f_1(x_1', x_2')| < \epsilon$ provided that (x_1', x_2') and (x_1'', x_2'') belong to $(-a - c \le x_1' \le a+c, -b-c \le x_2' \le b+c)$ and satisfy $|x_1'' - x_1'| < \delta(\epsilon)$ and $|x_2'' - x_2'| < \delta(\epsilon)$. Let

$$n_0 = n_0\epsilon = \text{Max}((c^3) + 1, (\delta^{-3}(\epsilon)) + 1),$$

and let $n > n_0$. Then $-a - c < x_1 - n^{-\frac{1}{3}} < x_1 + n^{-\frac{1}{3}} < a + c$, $-b - c < x_2 - n^{-\frac{1}{3}} < x_2 + n^{-\frac{1}{3}} < b + c$, and we have

$$(2) \quad \phi_n(x_1, x_2) = \pi^{-1} n \left[\int_{-a+c}^{a+c} dx_1' \int_{x_2+n^{-\frac{1}{3}}}^{b+c} dx_2' + \int_{-a-c}^{a+c} dx_1' \int_{x_2-n^{-\frac{1}{3}}}^{x_2+n^{-\frac{1}{3}}} dx_2' + \right.$$

$$\left. \int_{x_1-n^{-\frac{1}{3}}}^{x_1+n^{-\frac{1}{3}}} dx_1' \int_{x_2-n^{-\frac{1}{3}}}^{x_2+n^{-\frac{1}{3}}} dx_2' + \int_{x_1+n^{-\frac{1}{3}}}^{a+c} dx_1' \int_{x_2-n^{-\frac{1}{3}}}^{x_2+n^{-\frac{1}{3}}} dx_2' + \int_{-a-c}^{a+c} dx_1' \int_{-b-c}^{x_2-n^{-\frac{1}{3}}} dx_2' \right]$$

$$f_1(x_1', x_2') \exp \left[-n\{(x_1' - n_1)^2 + (x_2' - x_2)^2\} \right].$$

$$= T_1 + T_2 + \ldots + T_5,$$

say. In T_1 we have $|f_1(x_1', x_2')| < K$, $\exp[\ ] \le \exp(-n \cdot n^{-2/3})$, and so

$$(3) \quad |T_1| < \epsilon \, (n > n_1(\epsilon)).$$

Similarly[1]

$$(4) \quad |T_2|, |T_4|, |T_5| < \epsilon \quad (n > n_2(\epsilon)).$$

---

[1] I disclose at this point that in $T_2 = \int_{-a-c}^{a+c}$ is a 'slip' for $\int_{-a-c}^{x_1 - n^{-\frac{1}{3}}}$. One slip is practically certain in this style of writing, generally devil-inspired.

In T_3 write $x_1' = x_1 + x_1''$, $x_2' = x_2 + x_2''$. Since $|x_1''| \leq n^{-\frac{1}{3}}$, $|x_2''| \leq n^{-\frac{1}{3}}$ in the range concerned we have

(5) $\qquad |f_1(x_1 + x_1'', x_2 + x_2'') - f_1(x_1, x_2)| < \epsilon \, (n > \delta^{-3}(\epsilon))$.

Now in T_3 we have $f(x_1, x_2) = f_1(x_1, x_2)$. Hence

(6) $T_3 = T_{3,1} + T_{3,2}$, where

(7) $\quad T_{3,1} = \pi^{-1} n f(x_1, x_2) \int_{-n^{-\frac{1}{3}}}^{n^{-\frac{1}{3}}} dx_1'' \int_{-n^{-\frac{1}{3}}}^{n^{-\frac{1}{3}}} dx_2'' \exp\left[-n(x_1''^2 + x_2''^2)\right]$,

(8) $\qquad T_{3,2} = \pi^{-1} n \int_{-n^{-\frac{1}{3}}}^{n^{-\frac{1}{3}}} dx_1'' \int_{-n^{-\frac{1}{3}}}^{n^{-\frac{1}{3}}} dx_2'' \epsilon \{ f_1(x_1 + x_1'', x_2 + x_2'')$

$$-f_1(x_1, x_2)\} \times \exp\left[-n(x_1''^2 + x_2''^2)\right].$$

We have, for $n > \mathrm{Max}(n_0, n_1, n_2)$,

(9) $\qquad |T_{3,2}| \leq \pi^{-1} n \int_{-n^{-\frac{1}{3}}}^{n^{-\frac{1}{3}}} dx_1'' \int_{-n^{-\frac{1}{3}}}^{n^{-\frac{1}{3}}} dx_2'' \epsilon \exp\left[-n(x_1''^2 + x_2''^2)\right]$

$$\leq \pi^{-1} \epsilon n \int_{-\infty}^{\infty} dx_1'' \int_{-\infty}^{\infty} dx_2'' \exp\left[-n(x_1''^2 + x_2''^2)\right] = \epsilon.$$

Also the double integral in (7) is

(10) $\qquad \left(\int_{-n^{-\frac{1}{3}}}^{n^{-\frac{1}{3}}} e^{-nu^2} \, du \right)^2$

Now

$$\int_{-n^{-\frac{1}{3}}}^{n^{-\frac{1}{3}}} e^{-nu^2} \, du = 2 \int_0^{n^{-\frac{1}{3}}} = 2 \int_0^{\infty} - 2 \int_{n^{-\frac{1}{3}}}^{\infty}$$

$$= n^{-\frac{1}{2}} \pi^{1/2} - 2 \int_0^{\infty} e^{-n(n^{-\frac{1}{3}} + t)^2} \, dt$$

$$= n^{-\frac{1}{2}} \pi^{1/2} + O\left(e^{-n^{1/3}} \int_0^{\infty} e^{-2n^{2/3} t} \, dt \right)$$

$$= n^{-\frac{1}{2}} \pi^{1/2} \left(1 + O(n^{-\frac{1}{6}} e^{-n^{1/3}}) \right).$$

Hence it is easily seen that

$$\left|\left(\int\int_{-n^{-\frac{1}{3}}}^{n^{-\frac{1}{3}}} e^{-nu^2}\, du\right)^2 - n^{-1}\pi\right|$$

$$= \left|n^{-1}\pi\{1 + O(n^{-\frac{1}{6}}e^{-n^{1/3}})\} - n^{-1}\pi\right|$$

$$< \epsilon \quad (n > n_3(\epsilon)).$$

Hence from (10) and (7)

(11) $\qquad |T_{3,1} - f(x_1, x_2)| < K\epsilon \quad (n > \mathrm{Max}\,(n_0, n_1, n_2, n_3))\,.$

From (2) to (11) it follows that

$$|\phi_n(x_1, x_2) - f(x_1, x_2)| < K\epsilon \quad (n > \mathrm{Max}\,(n_0, n_1, n_2, n_3))\,,$$

and we have accordingly proved (1).

A civilized proof is as follows. Extend the definition of $f(x, y)$ to a larger rectangle R_+; e.g. on AB f is to be $f(A)$, and in the shaded square it is to be $f(C)$. The new f is continuous in R_+. Define, for (x, y) of R,

(i) $\qquad \phi_n(x, y) = \iint_{R_+} f(\xi, \eta) E\, d\xi d\eta \left/ \int_{-\infty}^{\infty}\int_{-\infty}^{\infty} E\, d\xi d\eta \right.,$

where $E = \exp[-n\{\xi - x)^2 + (\eta - y)^2\}]$. The denominator is the constant πn^{-1} (independent of x, y); hence (i) is equivalent to

(ii) $\qquad \phi_n(x, y) = \pi^{-1}n \iint_{R_+} f(\xi, \eta) E\, d\xi d\eta.$

The contributions to the numerator and to the denominator in (i) of (ξ, η)'s outside the square $S = S(x, y)$ of side $n^{-\frac{1}{3}}$ round (x, y) are exponentially small. The denominator itself being πn^{-1} we have (o's uniform)

$$\phi_n(x, y) = \left(\iint_S f(\xi, \eta) E\, d\xi d\eta \left/ \iint_S E\, d\xi d\eta \right.\right) + o(1).$$

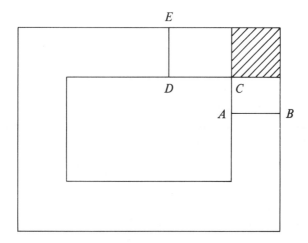

Figure 10

S being small, the $f(\xi, \eta)$ in the last numerator is $f(x,y)+o(1)$; so finally the M, defined by (ii) satisfies $\phi_n(x,y) = f(x,y) + o(1)$ as desired.

The second part of the proof of Weierstrass's theorem is as follows. For a suitable $N = N(n)$ we have, for all x, y of R and all ξ, η of R_+,

$$|E - \Sigma| < n^{-2},$$

where

$$\Sigma = \sum_{m=0}^{N} \frac{[-n\{(\xi - x)^2 + (\eta - y)^2\}]^m}{m!}$$

Then

$$\phi_n(x,y) = \Pi + o(1),$$

where $\Pi = \pi^{-1} n \iint_{R_+} \Sigma \, d\xi d\eta$, and is evidently a polynomial in (x,y).

Early writers had of course to work with what we should find intolerably clumsy tools and notations. An extreme case is a proof by Cauchy that every equation has a root. The modern version given in Hardy's *Pure Mathematics* (9th Edition, 494–496) could if necessary be telescoped to half a page. The ideas are all in Cauchy, but the reference to him has to be '*Exercises*, t. 4. 65–128': 64 pages (and all relevant, even though Cauchy is doing much pioneering by the way). The reading is not made any lighter by the fact that what we should call $\sum_0^n b''_m z^m$ has to appear as

$$(D''_0 + E''_0\sqrt{-1}) + (D''_1 + E''_1\sqrt{-1})(x + y\sqrt{-1}) + \ldots$$

$$+ (D''_n + E''_n\sqrt{-1})(x + y\sqrt{-1})^n.$$

Post-script on pictures

The 'pictorial' definition by Fig. 10, while the natural *source* of the idea, could in point of fact be given verbally with almost equal immediacy: 'Define f outside R to have the value at the nearest point of R;' this would be used in a printed paper if only to save expense, but the picture in a lecture. Here it serves as text for a sermon. My pupils *will* not use pictures, even unofficially and when there is no question of expense. This practice is increasing; I have lately discovered that it has existed for 30 years or more, and also why. A heavy warning used to be given[1] that pictures are not rigorous; this has never had its bluff called and has permanently frightened its victims into playing for safety. Some pictures, of course, are not rigorous, but I should say most are (and I use them whenever possible myself). An obviously legitimate case is to use a graph to define an awkward function (e.g. behaving differently in successive stretches): I recently had to plough through a definition quite comparable with the 'bad' one above, where a graph would have told the story in a matter of seconds. *This* sort of pictoriality does not differ in status from a convention like 'SW corner', now fully acclimatized. But pictorial *arguments*, while not so purely conventional, can be quite legitimate. Take the theorem[2]: '$f(x) = o(1)$ as $x \to \infty$ and $f'' = O(1)$ imply $f' = o(1)$'. If $f' \neq o(1)$

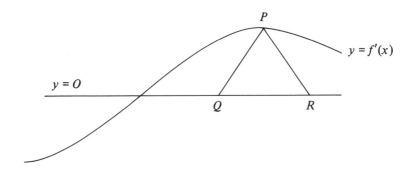

Figure 11

---

the graph $y = f'(x)$ will for some arbitrarily large x have 'peaks' (above or below $y = 0$) enclosing triangles like PQR, the height of P not small, the slopes of PQ, PR not large, and so the area PQR not small. Then $f(Q) - f(R)$ is not small, contradicting $f = o(1)$. This is rigorous (and printable), in the sense that in translating into symbols no step occurs that is not both unequivocal and trivial. For myself I *think* like this wherever the subject matter permits.

One of the best of pictorial arguments is a proof of the 'fixed point theorem' in one dimension: *Let $f(x)$ be continuous and increasing in $0 \leq x \leq 1$, with values satisfying $0 \leq f(x) \leq 1$, and let $f_2(x) = f\{f(x)\}$, $f_n(x) = f\{f_{n-1}(x)\}$. Then under iteration of f every point is either a fixed point, or else converges to a fixed point.*

For the professional the only proof needed is Fig. 12*

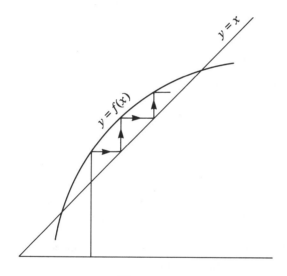

Figure 12

(Via Dr A. E. Western.) There was a Rent Act after 1914, and the definition of when a house was subject to it was as follows (my notation in brackets). The 'standard rent' (R) was defined to be the rent in 1914 (R_0), unless this was less than the rateable value (V), in which case it was to be the rateable value. 'The house is subject to the act if either the standard rent or the rateable value is less than £105.' There were many law suits, argued *ad hoc* in each case. The subject is governed by a fundamental theorem, unknown to the Law:

Theorem: *The house is subject to the act if and only if* $V < 105$.
This follows from[1]

Lemma. $\text{Min}\{\text{Max}(R_0, V), V\} = V$.

§2 *Misprints.* A recent number of the *Observatory* contained the charming 'typicle partical'.

Professor Offord and I recently committed ourselves to an odd mistake (*Annals of Mathematics* (2), 49, 923, 1.5). In formulating a proof a plus sign got omitted, becoming in effect a multiplication sign. The resulting false formula got accepted as a basis for the ensuing fallacious argument. (In defence, the final result was known to be true.)

In the *Mathematical Gazette* XIV, 23, there is a note on 'Noether's Canonical Curves', by W. P. Milne, in which he refers to 'a paper by Guyaelf (*Proc. London Math. Soc.*, Ser. 2, 21, part 5)'. Guyaelf is a ghost-author; the paper referred to is by W. P. Milne.

I once challenged Hardy to find a misprint on a certain page of a joint paper: he failed. It was in his own name: 'G, H. Hardy'.

A minute I wrote (about 1917) for the Ballistic Office ended with the sentence 'Thus σ should be made as small as possible'. This did not appear in the printed minute. But P. J. Grigg said, 'What is that?' A speck in a blank space at the end proved to be the tiniest σ I have ever seen (the printers must have scoured London for it).

In Music[2] a misprint can turn out to be a stroke of genius (perhaps the $A\sharp$ in bar 224 of the first movement of Beethoven's opus 106). Could this happen in Mathematics? I can imagine a hypothetical case. There was a time when infinite sets of intervals were unthought of, and a 'set of intervals' would be taken to mean a finite set. Suppose a definition of the 'content' of a set of points E, 'the lower bound of the total length of a set of intervals enclosing E'. Suppose now some precision, supplying the missing 'finite', should be moved to ask himself 'suppose an infinite set *were* allowed?' He would have set foot on a road leading inevitably to Lebesgue measure.

§3. *Verbalities.* I once came on a phrase like 'This will be true of the classes A, B, C provided C is not overlapped by A nor B.' The

---

[1] $\text{Min}(a, b)$ means the smaller, $\text{Max}(a, b)$ the larger, of a and b.
[2] Or Literature. 'Light falls from the (h)air'?

writer had queried the 'nor' but was told by a literary friend that it was necessary. It is obviously impossible mathematically: why? (The 'or' in 'A or B' is a mathematical symbol; 'A or B' is the mathematical name of a 'sum-class'.)

A recent (published) paper had near the beginning the passage 'The object of this paper is to prove (something very important).' It transpired[1] with great difficulty, and not till near the end, that the 'object' was an unachieved one.[2]

From an excellent book on Astronomy: 'Many of the spirals (galaxies), but very few of the ellipsoidals, show bright lines due, no doubt, to the presence or absence of gaseous nebulae.'

(This rich complex of horrors repays analysis. Roughly it is an illegitimate combination of the correct 'spirals show bright lines due to the presence ...' and the incorrect 'ellipsoidals don't show bright lines due to the absence ...'.)

The literary convention that numbers less than 10 should be given in words is often highly unsuitable in mathematics (though delicate distinctions are possible). The excessive use of the word forms is regrettably spreading at the present time. I lately came across (the lowest depth, from a very naïve writer) 'Functions never taking the values nought or one.' I myself favour using figures practically always (and am acting up to the principle in the book).

A linguist would be shocked to learn that if a set is not closed this does not mean that it is open, or again that 'E is dense in E' does not mean the same thing as 'E is dense in itself'.

'More than one is:' 'fewer than two are.'

'Where big X is very small.'

I considered including some paradoxical effects of the word 'nothing', but on consideration the thing is too easy.

The spoken word has dangers. A famous lecture was unintelligible to most of its audience because 'Hárnoo', clearly an important character in the drama, failed to be identified in time as h_ν.

---

[1] I have often thought that a good literary competition would be to compose a piece in which all the normal misuses of words and constructions were at first sight committed, but on consideration not.

[2] The author intended no dishonest claim, but his use of language was unusual.

I have had occasion to read aloud the phrase 'Where E' is any dashed (i.e. derived) set:' it is necessary to place the stress with care.

Jokes, etc.

§4. All the reflexive paradoxes are of course admirable jokes. Well-known as they are, I will begin with two classical ones.

(*a*) (Richard). There must exist (positive) integers that cannot be named in English by fewer than nineteen[1] syllables. Every collection of positive integers contains a least member, and there is a number N, 'the least integer not nameable in fewer than nineteen syllables'. But this phrase contains 18 syllables, and defines N.

(*b*) (Weyl). The vast majority of English adjectives do not *possess* the quality they *denote*; the adjective 'red' is not red: some, however, do possess it; e.g. the adjective 'adjectival'. Call the first kind heterological, the second homological: every adjective is one or the other. 'Heterological' is an adjective; which is it?

In a *Spectator* competition the following won a prize; subject: what would you most like to read on opening the morning paper?

OUR SECOND COMPETITION

The First Prize in the second of this year's competitions goes to Mr Arthur Robinson, whose witty entry was easily the best of those we received. His choice of what he would like to read on opening his paper was headed, 'Our Second Competition', and was as follows: 'The First Prize in the second of this year's competitions goes to Mr Arthur Robinson, whose witty entry was easily the best of those we received. His choice of what he would like to read on opening his paper was headed 'Our Second Competition', but owing to paper restrictions we cannot print all of it.'

Reflexiveness flickers delicately in and out of the latter part of Max Beerbohm's story *Enoch Soames*.

The following idea, a coda to the series, was invented too late (I do not remember by whom), but what *should* have happened is as follows. I wrote a paper for the *Comptes Rendus* which Prof. M. Riesz translated into French for me. At the end there were 3 footnotes. The first read (in

---

[1] Not '19', for sufficient if delicate reasons.

French) 'I am greatly indebted to Prof. Riesz for translating the present paper.' The second read 'I am indebted to Prof. Riesz for translating the preceding footnote.' The third read 'I am indebted to Prof. Riesz for translating the preceding footnote', with a suggestion of reflexiveness. Actually I stop legitimately at number 3: however little French I know I am capable of *copying* a French sentence.

Schoolmaster: 'Suppose x is the number of sheep in the problem.' Pupil: 'But, Sir, suppose x is not the number of sheep.' (I asked Prof. Wittgenstein was this not a profound philosophical joke, and he said it was.)

(A. S. Besicovitch) A mathematician's reputation rests on the number of bad proofs he has given. (Pioneer work is clumsy.)

'The surprising thing about this paper is that a man who *could* write it — would.'

'I should like to say how much this paper owes to Mr Smith.' 'Then why not do so?'

Tant pis, tante mieux: Auntie felt better. . . .

The most remarkable thing about Martin Luther was his diet of worms. He said: I can take no other course. (Via Dean Inge. A bad and made-up kind, but perhaps a supreme example.)

(Notice of religious service, 1938.) Subject: What shall I do in the crisis? Hymn: Search me, Oh Lord!

(A chestnut long ago, long period of oblivion, lately revived (1957).)

I once came upon a dignified person — e.g. headmaster, bishop — talking to his dog, all out: 'Ooo! My darlingest little dogsie wogsie...' What should (a) I, (b) he, have done? *I* should have said 'I do that too', but failed to rise to the occasion. What *he* did was to taper off: '*Good* doggie, good dog' (briskly).

'Honesty is the best policy.' Very well then; if I act so as to do the best for myself I am assured of acting honestly.

From time to time (since 1910) there were moves to get rid of the revolting optics and astronomy set in the Mathematical Tripos. It was discovered that over a period of years no wrangler attempted a question in either subject. An equivalent form of this proposition is that every attempt to do a question in optics or astronomy resulted in a failure to get a 1st.

'We all know that people can sometimes do better things than they have done, but — has done a better thing than he can do.' (An actual case, with agreement on the point among experts.)

An over-anxious research student was asking whether it was necessary to read all the literature before trying his hand. '*Nothing* is necessary — or sufficient.' The second part (embodying a harsh truth; the infinitely competent can be uncreative) arises inevitably by purely verbal association.

'Don't sniff at the sonatas of Archdukes, you never know who wrote them' (Haydn). (A propos of a rejected Ph.D. thesis.)

A too-persistent research student drove his supervisor to say 'Go away and work out the construction for a regular polygon of 65537 (= $2^{16} + 1$) sides.' The student returned 20 years later with a construction (deposited in the Archives at Göttingen).

A less painful story, which I certainly heard 25 or more years ago but will not vouch for the truth of, is that the first use of a crystal as a diffraction lattice was the result of taking seriously a suggestion made in jest. Such things could obviously happen. (I remember saying myself, on the theme that one should not import prejudices from daily life into, say, astrophysics: 'Be prepared to treat the sun as rigid or the interior of the earth as a perfect gas.' This was at a time when the stars were supposed to be at best very imperfect gases.)

'X finds gravitational waves in these conditions, but there is a suggestion that there is a mistake in the work.'

'Clearly *any* mistake generates gravitational waves.'

Landau kept a printed form for dealing with proofs of Fermat's last theorem. 'On page blank, lines blank to blank, you will find there is a mistake.' (Finding the mistake fell to the Privat Dozent.)

A precisian professor had the habit of saying: '... quartic polynomial $ax^4 + bx^3 + cx^2 + dx + e$, where e need not be the base of the natural logarithms.' (It might be.)

It was said of Jordan's writings that if he had 4 things on the same footing (as a, b, c, d) they would appear as a, M_3', ϵ_2, $\Pi_{1,2}''$.

'Liable to create a true impression.' (E.g. faking in an examination.)

'Less in this than meets the eye.'

Rock-climbing angles (*c.* 1900).

    Perpendicular — 60°.

    My dear Sir, absolutely perpendicular — 65°.

    Overhanging — 70°.

I read in the proof-sheets of Hardy on Ramanujan: 'As someone said, each of the positive integers was one of his personal friends.' My reaction was, 'I wonder who said that; I wish I had.' In the next proof-sheets I read (what now stands), 'It was Littlewood who said...'

(What had happened was that Hardy had received the remark in silence and with poker face, and I wrote it off as a dud. I later taxed Hardy with this habit; on which he replied: 'Well, what is one to do, is one always to be saying "damned good"?' To which the answer is 'yes'.)

I end with the joke of my own that gives me most pleasure to recall. Veblen was giving a course of 3 lectures on 'Geometry of Paths'. At the end of one lecture the paths had miraculously worked themselves into the form

$$\frac{x-a}{l} = \frac{y-b}{m} = \frac{z-c}{n} = \frac{t-d}{p}.$$

He then broke off to make an announcement about what was to follow, ending with the words 'I am acting as my own John the Baptist.' With what meaning I do not now recall (certainly not mine), but I was able to seize the Heaven-sent opportunity of saying 'Having made your own paths straight.'

*THE ZOO

§1. The domain obtained by removing an infinity of shaded sectors as in Fig. 13 has very important applications in function-theory (too high-brow to mention here). It is commonly called the amoeba or the star-fish domain.

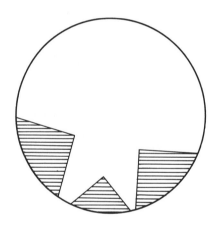

Figure 13

The snake. Representing the domain shown in Fig. 14 on a unit circle we have a function $f(z)$ that takes some values twice (those in the twice covered region), but for which f' never vanishes. (The fallacy that f' must vanish is absurdly common — doubtless an effect of too steady a diet of algebraic function-theory, in which all sheets of the Riemann surfaces are alike and extend over the whole plane.)

Figure 14

The crocodile (Fig. 15). The teeth overlap and have a total length just infinite. If the domain is represented on a unit circle we have an example of a function $f(z) = \sum c_n z^n$ in which the real part $U(\theta)$ of $f(e^{i\theta})$ is of bounded variation, and the imaginary part $V(\theta)$ is as nearly so as we like. On the other hand

Figure 15

$$\sum |c_n| = \int_0^1 \left(\sum n|c_n|\rho^{n-1}\right) d\rho \geq \int_0^1 |f'(\rho)| d\rho,$$

and the last integral is the length of the image of the radius vector $(0,1)$ of the z-circle. This image, however, is some path winding between the teeth to the nose and has infinite length. Hence $\sum |c_n|$ is divergent.

If both U and V are of bounded variation it is a known theorem that the series is convergent. The crocodile shows that the result is best possible, a question I had been asked (by Prof. L. C. Young) to decide. When returning to Cambridge along the Coton footpath the 'hippopotamus' (Fig. 16, a well-known[1] character in the theory of 'prime-ends', but only now baptised in imitation of the crocodile) flashed into my mind from nowhere. He did not quite do the trick (or so I thought), but a couple of hundred yards on he switched to a crocodile.

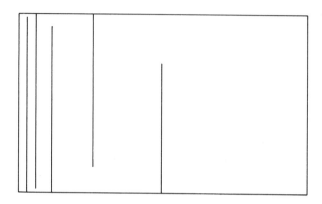

Figure 16

The hedgehog. Suppose a topological transformation T is such that for each point P of the plane $T^n P$ *ultimately* (i.e. for $n > n_0(P)$) gets into a bounded domain a *and stays there.* Let Δ_+ be Δ slightly enlarged (i.e. the closure $\overline{\Delta}$ of Δ is contained in Δ_+). Let $\overline{D}$ be any closed bounded domain. Is it the case that $T^n P$ converges *uniformly* into Δ_+ for all P of $\overline{D}$ (i.e. $T^n \overline{D} \subset \Delta_+$ for $n > n_0(\overline{D})$)? Everyone's first guess is yes (and the corresponding thing is true in 1 dimension), but the answer is in fact no. For this Miss Cartwright and I found the example of Fig. 17. (There are an infinity of spines running to L, L' as limit points.) Consider a T which leaves the hedgehog (the figure of full lines) invariant as a whole, but transforms each spine into the next one to the right, and further imposes a general contraction of outside points towards the boundary of the hedgehog. Δ is the domain bounded by the dotted line, Δ_+ is Δ arbitrarily little enlarged. While $T^n P$ is ultimately inside Δ for each P, the tip of a spine near L requires a large n for $T^n P$ to get finally back into Δ_+.

We later found a much simpler example, Fig. 18, in which u, s, c are respectively totally unstable, stable, and saddle-point (col) fixed points

---

[1]So well-known, in fact, that my artist does not feel he can take liberties with him.

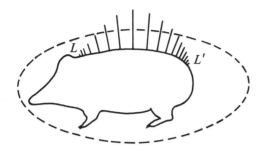

Figure 17

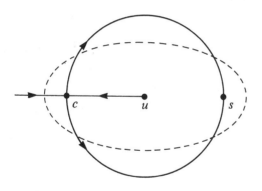

Figure 18

of the T, and the lines of the figure are invariant, as wholes, under T. Δ is the domain enclosed by the dotted line. For any P $T^n P$ ultimately stays in Δ, but points near u, or again points near the line uc, take arbitrarily long over ending up in Δ_+. This T however, leaves a whole area invariant, and the example does not cover the important class of T's for which every bounded area shrinks to zero area under iteration of T. The hedgehog does cover this if his body area is reduced to zero; he is not debunked, only disinflated.*

BALLISTICS

§1. '*The rifleman's problem.*' Should the t.e ϕ (tangent elevation, i.e. elevation above the line of sight of a target) for a given range be increased or decreased if the target is slightly above the horizontal? The answer is probably not intuitive; but it *is* intuitive that there is a decrease for a target below the horizontal, so the rate of increase with the angle of sight α is presumably positive, and we infer an increase for the original question. For moderate ranges ϕ tends to 0 as α tends to $\frac{1}{2}\pi$, so the initial increase later becomes a decrease. The upshot is that it is a reasonably good approximation to keep ϕ constant for all small positive α[1]; this principle is called 'the rigidity of the trajectory'.

The professional (rifleman) believes in the initial increase; he feels unhappy in firing uphill (and happy firing down). 'The bullet has to pull against the collar.' It is arguable that he is right; on the one hand the pull is there; on the other the correction that leads ultimately to a decrease is only second order near the horizontal.

In vacuo the height H of a trajectory with horizontal target is $\frac{1}{8}gT^2(= 4T^2)$, where T is the time of flight. This happens to be a pretty good approximation over all sorts of guns and all sorts of elevations (even vertical ones).

Suppose we accept these two principles as absolute instead of approximate. Then a curiously ingenious argument becomes possible to arrive at the *position* of the vertex of a trajectory, given only the range table. *The range table gives in effect any two of R (range), ϕ (elevation), and T (time of flight, with $H = 4T^2$ linked with it) as functions

---

[1]The approximation is improved by the diminishing density of the air upwards. Details are easily worked out for trajectories *in vacuo*; the *relative* behaviour of actual ones is not very different (and in any case tends to the *in vacuo* behaviour as $\phi \to 0$).

of the third. In Fig. 19 we have, for a given ϕ,

$$\theta + \alpha = \phi, \quad r = R(\theta), \quad H = 4T^2, \quad \sin \alpha = H/r,$$

whence

$$R(\theta) \sin(\phi - \theta) = 4T^2(\phi),$$

from which a root $\theta(\phi)$ can be found by trial and error, and thence α and ON.*

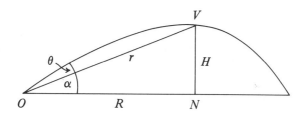

Figure 19

My personal contact with this was odd. On a night not long before I went to the Ballistic Office (about Dec., 1915) I was orderly officer for a large Artillery camp with many senior officers (but no professional ballisticians). Lying on the office table was the above figure and enough symbols to show what it was about. A day or so afterwards a Colonel asked me whether it was possible to find the position of V. I gravely reproduced the argument, and as it was new to him it amused me to say no more. (I never discovered its source.)

§2. *Rockets.* The trajectory of a particle under gravity and a constant force is an old toy of particle dynamics. Theoretically the trajectory cannot be started from rest except vertically. (With a non-vertical start the initial curvature is infinite.) What curves have elegant (*a*) shapes, (*b*) equations? A bomb-trajectory has approximately an equation

$$e^{-y} = k \cos x.$$

I heard an account of the battle of the Falkland Islands (early in the 1914 war) from an officer who was there. The German ships were destroyed at extreme range, but it took a long time and salvos were continually falling 100 yards to the left. The effect of the rotation of the earth is similar to 'drift' and was similarly incorporated in the gun-sights. But this involved the tacit assumption that Naval battles take place round about latitude 50° N. The double difference for 50° S and extreme range is of the order of 100 yards.

*§3.   Suppose a particle is projected vertically downwards in a medium whose density ρ increases with the depth y like $1/(1 - \lambda y)$, and whose resistance varies as $\rho \nu^2$, so that the deceleration due to air resistance is $1/(1 - \lambda y)$. If now we have $\mu = \lambda$, then, whatever the initial velocity, *the motion is simple harmonic* (so long as it lasts; the bottom end of the amplitude occurs where the density becomes infinite). Various attempts of mine to set this in examinations failed. I had hoped to draw the criticism of 'unreality', to which there is the following reply. In 1917-18, a range table was called for, for the first time, and quickly, for a gun in an aeroplane flying at a fixed height, to fire in all directions. A method existed, based on numerical calculation of the vertically upward and vertically downward trajectories. It happened that within the permissible limits of accuracy the values of λ and μ could be faked to make $\lambda = \mu$ (and $\rho = (1 - \lambda y)^{-1}$ was a sufficiently accurate density law). The downward trajectory could accordingly be read off from a table of sines, and the range table was in fact made in this way (in about two-thirds the time it would otherwise have taken).

§4. I do not deny that the example just given is slightly disreputable; here is a more respectable one. For any 'property' of a trajectory in an atmosphere of varying density, say the property of having range R for given elevation ϕ (and fixed 'gun'[1]), there is an 'equivalent homogeneous atmosphere' (in which the R for the given ϕ is the same as in the actual atmosphere; it varies of course with ϕ); this is expressed by a fraction c, the equivalent constant density being that at height ch, where h is the greatest height in the trajectory. Now it is always the case, in any such problem, that in the limit as the length of the trajectory tends to 0, c is a pure number, independent of the law of resistance and the rate of variation of density. In the particular problem referred to the limiting value is $c = \frac{3}{5}$. (c varies with the problem and is, e.g. $\frac{2}{5}$ for 'time of flight given ϕ'. To establish these results from first principles requires rather heavy calculations, but these can be eased by using the fact that the limit must be a pure number.) The 'average height' in any ordinary sense being $\frac{2}{3}h$, this is a mild subtlety (the Office would not believe it

---

[1] A 'gun' is an ordered pair of constants, (C, V).

until they had made a numerical experiment, after which they believed any result guaranteed by theory).

It is as intuitive as anything can be that, whatever the 'property', c cannot be outside the range $(0,1)$: how does the reader react to the possibility of $c = 0$ or 1?

As a matter of fact there exists a very simple (and practically important) case in which $c = 0$: the problem is 'time of flight on a given inclined plane given the elevation'.

The height h being here only first order (in the flat trajectory it is second order), the calculations are simpler. If we use the principle that c is independent of the law of resistance and the rate of variation (the paradox is in any case fully alive in the limiting case) we can simplify as follows. With gravity and initial velocity normalized to 1 we may suppose the retardation to be of the form $\mu(1 - \lambda y)\nu$, *where ν is small*. Let ϕ be the t.e., α the angle of the inclined plane (Fig. 20). The actual trajectory is the solution of

$$\ddot{x} = -\mu(1 - \lambda y)\dot{x}, \quad \ddot{y} = -1 - \mu(1 - \lambda y)\dot{y},$$

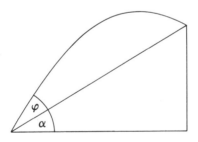

Figure 20

with initial values $\dot{x}_0 = \cos(\phi + \alpha)$, $\dot{y}_0 = \sin(\phi + \alpha)$. The time τ at which $y = x \tan \alpha$ is to be equated to the corresponding time with $\lambda = 0$ and $\mu' = (1 - ch)\mu$ for μ. We are to take the limit as ϕ (or τ) and μ tend to 0 ($\lambda \to 0$ is not necessary); this means *inter alia* that we can ignore all terms involving μ^2. Then approximately

$$x^{(3)} = \lambda\mu\dot{x}\dot{y}, x^{(4)} = -\lambda\mu x, x^{(5)} = 0;$$
$$y^{(3)} = \mu(1 - \lambda y) + \lambda\mu\dot{y}^2, y^{(4)} = -\lambda\mu\dot{y} + 2\lambda\mu\dot{y}\ddot{y} = -3\lambda\mu\dot{y}, y^{(5)} = 0.$$

At time τ we have, writing γ, σ for $\cos(\phi + \alpha)$, $\sin(\phi + \alpha)$,

$$\gamma \tan \alpha = \frac{y}{x/\gamma} = \sum_1^4 y_0^{(n)} \frac{\tau^n}{n!} \bigg/ \sum_1^4 \frac{x_0^{(n)}}{\gamma} \frac{\tau^n}{n!}$$

$$= \frac{\sigma - \frac{1}{2}(1 + \mu\sigma)\tau + \frac{1}{6}(\lambda\mu\sigma^2 + \mu)\tau^2 - \frac{3}{24}\lambda\mu\sigma\tau^3}{1 - \frac{1}{2}\mu\tau + \frac{1}{6}\lambda\mu\sigma\tau^2 - \frac{1}{24}\lambda\mu\tau^3}$$

$$= \sigma - \frac{1}{2}\tau - \frac{1}{12}\mu\tau^2 + O(\tau^4),$$

by straightforward calculation, ignoring μ^2: note that there is no term in τ^3.

The right hand side is to be equal to

$$\sigma - \frac{1}{2}\tau - \frac{1}{12}\mu(1 - c\sigma\tau)\tau^2 + O(\tau^4);$$

hence $c = O(\tau)$, and $c = 0$ in the limit.*

THE DILEMMA OF PROBABILITY THEORY

There is a solid body of propositions of the theory, and no one dreams of doubting their practical applicability. If, for example, a card is drawn 1300 times at random from a (whole) ordinary pack we should be surprised if the number of aces differed greatly from 100, and we believe the more refined statements that it is about an even chance that the number will lie between 94 and 106 inclusive, and that it is one of less than 1 in 10^6 that it will lie outside the range 50 to 150. To avoid possible misunderstanding I begin with a certain distinction. 'The probability of drawing an ace is $\frac{1}{13}$; the probability of drawing an ace twice running is $\left(\frac{1}{13}\right)^2$.' Such statements, and most of the 'probability' one meets in algebra text-books, are effectively pure mathematics; the underlying conventions about 'equal likelihood' are automatic, and the subject matter reduces to 'permutations and combinations'. This side of probability theory will not concern us.

The earlier statements are very different; they assert about the real world that such and such events will happen with such and such a probability; they intend this in the commonsense meaning, and do not intend to say, e.g. that of the 52^{1300} ways of drawing 1300 cards a certain proportion (near $\frac{1}{2}$) contain from 94 to 106 aces. The question now is about the foundations of the subject.

Mathematics (by which I shall mean pure mathematics) has no grip on the real world; if probability is to deal with the real world it must contain elements outside mathematics; the *meaning* of 'probability' must relate to the real world, and there must be one or more 'primitive' propositions about the real world, from which we can then proceed deductively (i.e. mathematically). We will suppose (as we may by lumping several primitive propositions together) that there is just one primitive proposition, the 'probability axiom', and we will call it A for short. Although it has got to be *true*, A is by the nature of the case incapable of deductive proof, for the sufficient reason that it is about the real world (other sorts of 'justification' I shall consider later).

There are 2 schools. One, which I will call mathematical, stays inside mathematics, with results that I shall consider later. We will begin with the other school, which I will call philosophical. This attacks directly the 'real' probability problem; what are the axiom A and the meaning of 'probability' to be, and how can we justify A? It will be instructive to consider the attempt called the 'frequency theory'. It is natural to believe that if (with the natural reservations) an act like throwing a die is repeated n times the proportion of 6's will, *with certainty*, tend to a limit, p say, as $n \to \infty$. (Attempts are made to sublimate the limit into some Pickwickian sense—'limit' in inverted commas. But either you *mean* the ordinary limit, or else you have the problem of explaining how 'limit' behaves, and you are no further. You do not make an illegitimate conception legitimate by putting it into inverted commas.) If we take this proposition as 'A' we can at least settle off-hand the other problem, of the *meaning* of probability; we define its measure for the event in question to be the number p. But for the rest this A takes us nowhere. Suppose we throw 1000 times and wish to know what to expect. Is 1000 large enough for the convergence to have got under way, and how far? A does not say. We have, then, to add to it something about the rate of convergence. Now an A cannot assert a *certainty* about a particular number n of throws, such as 'the proportion of 6's will *certainly* be within $p \pm \epsilon$ for large enough n (the largeness depending on ϵ)'. It can only say 'the proportion will lie between $p \pm \epsilon$ *with at least such and such probability (depending on ϵ and n_0) whenever $n > n_0$*'. The vicious circle is apparent. We have not merely failed to *justify* a workable A; we have failed even to *state* one which would work if its truth were granted. It is generally agreed that the frequency theory won't work. But whatever the theory it is clear that the vicious circle is very deep-seated: certainty being impossible, whatever A is made to state can be stated only in terms of 'probability'. One is tempted to the extreme rashness of saying that the problem is insoluble (within our current conceptions). More sophisticated attempts than the frequency theory have been made, but they fail in the same sort of way.

I said above that an A is inherently incapable of deductive proof. But it is also incapable of inductive proof. If inductive evidence is offered 'in support' we have only to ask *why* it supports (i.e. gives probability to) A. Justification of a proposition (as opposed to an axiom) can be given only in terms of an earlier proposition or else of an axiom; justification of a *first* proposition, therefore, only in terms of an axiom. Now any answer to the question 'why' above is a 'first' proposition; but the only axiom there is to appeal to is A itself (or part of it), and it is A we are trying to justify. So much for the philosophical school.

The mathematical school develops the theory of a universe of ideal 'events' E and a function $p(E)$ which has the E's as arguments. Postulates[1] are made about the E's and the function p; unlike an axiom A, these are not true or false (or even meaningful), but are strictly parallel to the 'axioms' of modern geometry. The development of the logical consequence of the postulates is a branch of pure mathematics, though the postulates are naturally designed to yield a 'model' of the accepted body of probability theory. This is in many ways a desirable development: the postulates are chosen to be a minimal set yielding the model theory, and any philosophical discussion can concentrate on them.

Some of the remoter parts of the ordinary theory (e.g. inverse probability) are philosophically controversial; these can be separated from the rest in the model by a corresponding separation of postulates. The purely technical influence of the method on the ordinary theory is also far from negligible; this is a usual result in mathematics of 'axiomatizing' a subject. (Incidentally the most natural technical approach is to work with quite general 'additive sets', with the result that the aspiring reader finds he is expected to know the theory of the Lebesgue integral. He is sometimes shocked at this, but it is entirely natural.)

We come finally, however, to the relation of the ideal theory to the real world, or 'real' probability. If he is consistent a man of the mathematical school washes his hands of applications. To someone who wants them he would say that the ideal system runs parallel to the usual theory: 'If this is what you want, try it: it is not my business to justify application of the system; that can only be done by philosophizing; I am a mathematician'. In practice he is apt to say: 'try this; if it works that will justify it'. But now he is not merely philosophizing; he is committing the characteristic fallacy. Inductive experience that the system works is not evidence.

---

[1] These are generally called 'axioms', but I am using 'axiom' in another sense.

FROM FERMAT'S LAST THEOREM TO THE
ABOLITION OF CAPITAL PUNISHMENT[1]

It is a platitude that pure mathematics can have unexpected consequences and affect even daily life. Could there be a chain of ideas such as the title suggests? I think so, with some give and take; I propose to imagine at one or two points slight accidental changes in the course of mathematical history. The amateur should perhaps be warned that the thesis takes some time to get under way, but moves rapidly at the end; I hope he may be persuaded to stay the earlier part of the course (which incidentally is concerned with ideas of great mathematical importance).

The theory of numbers is particularly liable to the accusation that some of its problems are the wrong sort of questions to ask. I do not myself think the danger is serious; either a reasonable amount of concentration leads to new ideas or methods of obvious interest, or else one just leaves the problem alone. 'Perfect numbers' certainly never did any good, but then they never did any particular harm. F.L.T. is a provocative case; it bears every outward sign of a 'wrong question' (and is a negative theorem at that); yet work on it, as we know, led to the important mathematical conception of 'ideals'. This is the first link in my chain of ideas.

The intensive study of F.L.T. soon revealed that to gain deeper insight it is necessary to *generalize* the theorem[2] the x, y, z of the 'impossible' $x^p + y^p = z^p$ were generalized from being ordinary integers to being integers of the 'field' of the equation $\varsigma^p + 1 = 0$. If α is a root (other than -1) of this equation, then the integers of the field are, nearly enough for present purposes, all the numbers of the form $m_0 + m_1\alpha + \ldots + m_{p-2}\alpha^{p-1}$, where the m's are 'ordinary' integers (of

---

[1] I gave the substance of this in a paper at Liverpool about 1929. F.L.T. asserts that for an integer n greater than 2 the equation $x^n + y^n = z^n$ is impossible in integers x, y, z all different from 0. It is enough to settle the special case in which n is a prime p. Its truth remains undecided.

[2] And so to attack an apparently more difficult problem!

either sign). The idea of the divisibility of a field integer a by another, b, is simple enough; a is divisible by b if $a = bc$, where c is a field integer. Again, a field prime is a field integer with no 'proper' divisor, that is, is divisible only by itself and by the field 'unities' (generalizations of '1', they divide all field integers). Any (field) integer can further be resolved into prime 'factors'. But now a new situation develops with the fields of some (indeed most) p's, resolution into prime factors is not (as it is for ordinary integers) always unique. 'Ideals' are called in to restore uniqueness of factorization.[1]

New entities like ideals generally begin as a 'postulation', being later put on a rigorous basis by the 'construction' of an entity which behaves as desired.[2] The easiest course at this point is to give at once Dedekind's construction, and go on from there. If $\alpha, \beta, \ldots, \kappa$ are any finite set of field integers, consider the class of *all* numbers (they are field integers) of the form $m\alpha + \ldots + k\kappa$, where $m, \ldots, k$ are ordinary integers; a number of the class is 'counted only once' if there is overlapping. The class, which is completely determined by the set $\alpha, \ldots, \kappa$, is denoted by $(\alpha, \ldots, \kappa)$ and is called an 'ideal'. Let us now go back to the 'field' of ordinary integers and see what an 'ideal' becomes in that special case. The (ordinary) integers $\alpha, \ldots, \kappa$ have a 'greatest common divisor' d (and this fact is the basis of 'unique factorization' — Euclid makes it so in his proof, and this is the 'right' proof, though text-books often give another). The class of numbers $m\alpha + \ldots + k\kappa$, when the extensive overlapping is ignored, is easily seen to be identical with the class of numbers nd (n taking all ordinary integral values); the ideal $(\alpha, \ldots, \kappa)$ is identical with the ideal (d). An ideal in the general field which is of the form (α) (α being a field integer) is called a 'principal ideal': the field of ordinary integers has, then, the property that *all* its ideals are principal. Suppose next that a, b are ordinary integers and that b divides a, e.g. let $a = 6, b = 3$. Then (a) is the class of all multiples of 6, (b) the class of all multiples of 3, and *the class (a) is contained in the class (b).* Conversely this can happen only if a *is* divisible by b. Thus 'b divides a' and '(a) is contained in (b)' are exactly equivalent. Now the set of entities (a) is in exact correspondence with the set of entities a (without the brackets); we can take the ideals (a) for raw material instead of the integers a, and interpret '(b) divides (a)' as meaning '(a) is contained in (b)'. The theory of the bracketed entities runs parallel to that of the unbracketed ones, and is a mere 'translation' of the latter. Return to the general field. The integer α gets replaced by (a), but all ideals are no longer principal; the totality of *all* ideals is taken for the raw material, and divisibility of a 1st ideal by a 2nd (so far not defined) is taken to mean the containing of the 1st ideal

---

[1] For integers of the field of a general algebraic equation $a_0 + a_1\varsigma + \ldots + a_n\varsigma^n = 0$, where the a's are ordinary integers.

[2] Other instances: complex numbers, points at infinity; non-Euclidean geometry.

(qua class) in the 2nd. Suppose now, denoting ideals by Clarendon type, that $a, \ldots, k$ are a finite set of ideals. There is then an ideal d whose class contains each of the classes of $a, \ldots, k$, and which is the smallest of this kind[1]: d functions as a 'greatest common divisor' of $a, \ldots, k$. After this we arrive without difficulty (and much as in the 'ordinary' case) at the key proposition that every ideal can be factorized uniquely into 'prime ideals'. Since this theory 'reduces' to the 'ordinary' theory in the special case of 'ordinary' integers it is a genuine *generalization* of the latter, and may legitimately be said to 'restore' unique factorization.

I feel that ideals 'ought' to have been created first, and to have suggested the famous 'Dedekind section' definition of 'real numbers', but though it was a near thing the facts are otherwise.[2] We will, however, suppose history modified.

In a Dedekind section *all* rational numbers fall into one of two classes, L and R,[3] every member of L being to the left of (i.e. less than) every member of R (and for definiteness L has no greatest member — R may or may not happen to have a least one). The totality of all possible 'sections of the rationals' provides a set of entities with the properties we wish the continuum of 'real numbers' to have, and real numbers become properly founded.

What exactly does 'section' ('Schnitt') mean? After the class definition of the ideal it would seem natural, almost inevitable, to define it, and the real number also, to *be* the class L (of course X would do equally well). Thus the real number $\sqrt{2}$ *is* the class of rationals r composed of the negative ones together with the non-negative ones satisfying $r^2 < 2$. It is reasonable to take the step for granted and call this Dedekind's definition. The actual circumstances are very strange. For Dedekind the Schnitt is an act of cutting, not the thing cut off; he 'postulates' a 'real number' to do the cutting and is not entirely happy about it (and the modern student is much happier with the *class*): as Bertrand Russell says, the method of postulation has many advantages, which are the same as those of theft over honest toil. Incidentally, on a point of linguistics, both 'Schnitt' and 'section' are ambiguous and can mean either the act of cutting or the thing cut off: it is a case in which a misreading

---

[1]If $a = (\alpha_1, \beta_1, \ldots, \kappa_1), \ldots k = (\alpha_n, \beta_n, \ldots, \kappa_n)$, then actually $d = (\alpha_1, \ldots, \kappa_1, \ldots, \alpha_n, \ldots, \kappa_n)$.

[2]Publication was more or less contemporaneous (and the later idea was available for revision of the earlier), but the 'section', published 1872 ('Was sind usw.?'), originated in 1858.

[3]The letters L, R, for which a generation of students is rightly grateful, were introduced by me. In the first edition of *Pure Mathematics* they are T, U. The latest editions have handsome references to me, but when I told Hardy he should acknowledge this contribution (which he had forgotten) he refused on the ground that it would be insulting to mention anything so minor. (The familiar response of the oppressor: what the victim wants is not in his own best interests.)

could have constituted an advance.

These two 'class' definitions (ideal and real number) have no parallel since about 350 B.C. Eudoxus's (Fifth book of Euclid) definition of 'equal ratio' (of incommensurables) is in fact very near the Dedekind section (Eudoxus's equal $a : b$ and $c : d$ correspond each to the same 'class of rationals m/n; the two ratios are to be equal if the class of m/n's for which $ma < nb$ is identical with that for which $mc < nd$'.)

Turn now to another question: what is meant by a 'function'? I will digress (though with a purpose) to give some extracts from Forsyth's *Theory of Functions of a Complex Variable*; this is intended to make things easy for the beginner. (It was out of date when written (1893), but this is the sort of thing my generation had to go through. The fact that 'regularity' of a function of a complex variable is being explained at the same time adds unfairly to the general horror, but I should be sorry to deprive my readers of an intellectual treat.)

> All ordinary operations effected on a complex variable lead, as already remarked, to other complex variables; and any definite quantity, thus obtained by operations on z, is necessarily a function of z.

> But if a complex variable w is given as a complex function of x and y without any indication of its source, the question as to whether w is or is not a function of z requires a consideration of the general idea of functionality.

> It is convenient to postulate $u + iv$ as a form of the complex variable, where u and v are real. Since w is initially unrestricted in variation, we may so far regard the quantities u and v as independent and therefore as any functions of x and y, the elements involved in z. But more explicit expressions for these functions are neither assigned nor supposed.

> The earliest occurrence of the idea of functionality is in connection with functions of real variables; and then it is coextensive with the idea of dependence. Thus, if the value of X depends on that of x and on no other variable magnitude, it is customary to regard X as a function of x; and there is usually an implication that X is derived from x by some series of operations.

> A detailed knowledge of z determines x and y uniquely; hence the values of u and v may be considered as known and therefore also w. Thus the value of w is dependent on that of z, and is independent of the values of variables unconnected with z; therefore, with the foregoing view of functionality, w is a function of z.

It is, however, consistent with that view to regard as a complex function of the two independent elements from which z is constituted; and we are then led merely to the consideration of functions of two real independent variables with (possibly) imaginary coefficients.

Both of these aspects of the dependence of w on z require that z be regarded as a composite quantity involving two independent elements which can be considered separately. Our purpose, however, is to regard z as the most general form of algebraic variable and therefore as an irresoluble entity; so that, as this preliminary requirement in regard to z is unsatisfied, neither of these aspects can be adopted.

Suppose that w is regarded as a function of z in the sense that it can be constructed by definite operations on z regarded as an irresoluble magnitude, the quantities u and v arising subsequently to these operations by the separation of the real and imaginary parts when z is replaced by $x + iy$. It is thereby assumed that one series of operations is sufficient for the simultaneous construction of u and v, instead of one series for u and another series for v as in the general case of a complex function [above]. If this assumption be justified by the same forms resulting from the two different methods of construction, it follows that the two series of operations, which lead in the general case to u and to v, must be equivalent to the single series and must therefore be connected by conditions; that is, u and v as functions of x and y must have their functional forms related:

$$(1) \qquad \frac{\partial w}{\partial x} = \frac{1}{i} \frac{\partial w}{\partial y} = \frac{dw}{dy}$$

$$(2) \qquad -\frac{\partial v}{\partial x} = \frac{\partial u}{\partial y}, \quad \frac{\partial u}{\partial x} = \frac{\partial v}{\partial y}.$$

These are necessary ... and sufficient ... relations between the functional forms of u and v.

The preceding determination of the necessary and sufficient conditions of functional dependence is based on the existence of a functional form; and yet that form is not essential, for, as already remarked, it disappears from the equations of condition. Now the postulation of such a form is equivalent to an assumption that the function can be numerically calculated for each particular value of the independent variable, though the immediate expression of the assumption has disappeared in the present case. Experience of functions of real variables shews that it is often more convenient to use their properties than to possess their numerical values. This experience is confirmed by what has preceded. The essential conditions of functional dependence are the equations (1) ...

Nowadays, of course, a function $y = y(x)$ means that there is a class of 'arguments' x, and to each x there is assigned 1 and only 1 'value' y. After some trivial explanations (or none?) we can be balder still, and say that a function is a class C of pairs (x, y) (order within the bracket counting), C being subject (only) to the condition that the x's of different pairs are different. (And a 'relation' R, 'x has the relation R to y', reduces *simply* to a class, which may be any class whatever, of ordered pairs.) Nowadays, again, the x's may be any sort of entities whatever, and so may the y's (e.g. classes, propositions). If we *want* to consider well-behaved functions, e.g. 'continuous' ones of a real variable, or Forsyth's $f(z)$, we *define* what being such a function means (2 lines for Forsyth's function), and 'consider' the class of functions so restricted. That is all. This clear daylight is now a matter of course, but it replaces an obscurity as of midnight.[1] The main step was taken by Dirichlet in 1837 (for functions of a real variable, the argument class consisting of some or all real numbers and the value class confined to real numbers). The complete emancipation of e.g. propositional functions belongs to the 1920's.

Suppose now, again to imagine a modified history, that the way out into daylight had been slightly delayed and pointed (as it so easily might have been) by the success of Dedekind's ideas. I will treat the idea of function, then, as derived from the Fermat theorem. (If this is rejected 'abolition' will be related instead to Fourier series or the differential equations of heat conduction.)

Consider now a function in which the argument class consists of the moments t of (historical) time and the value $f(t)$ for argument t is a state of the Universe (described in sufficient detail to record any happening of interest to anybody). If t_0 is the present date, $f(t)$, for $t < t_0$ is a description, or dictionary, of what *has* happened. Suppose now the dictionary transported back to an earlier time τ; then it contains a prediction of what is going to happen between times τ and t_0 This argument is clearly *relevant* to the issue of determination versus free-will and could reinforce any existing doubts. Doubts about free-will bear on the problem of moral responsibility and so (rightly or wrongly) on the problem of punishment. Wilder ideas have influenced reformers.

---

[1]The trouble was, of course, an obstinate feeling at the back of the mind that the value of a function 'ought' to be got from the argument by 'a series of operations'.

A MATHEMATICAL EDUCATION

It is my education. It illustrates conditions before 1907, but has some oddities of its own.

I am sure that I do not suffer from the weakness of false modesty, and to begin with I do not mind saying that I was precocious: as a matter of fact precocity in a mathematician has no particular significance one way or the other, and there are plenty of examples both ways; I happen to belong to the precocious class.

Born June 9, 1885, I was in South Africa from 1892 to 1900; I left the Cape University at the age of 14, and after 2 or 3 months went to England to go to St. Paul's School, where I was taught for 3 years by F. S. Macaulay. My knowledge then was slight by modern standards; the first 6 books of Euclid, a little algebra, trigonometry up to solution of triangles. During my 3 years at St. Paul's I worked intensively; seriously overworked indeed, partly because it was a period of severe mental depression.

The tradition of teaching (derived ultimately from Cambridge) was to study 'lower' methods intensively before going on to 'higher' ones; thus analytical methods in geometry were taken late, and calculus very late. And each book was more or less finished before we went on to the next. The accepted sequence of books was: Smith's *Algebra*; Loney's *Trigonometry*; *Geometrical Conics* (in a very stiff book of Macaulay's own: metrical properties of the parabola, for instance, gave scope for infinite virtuosity); Loney's *Statics and Dynamics*, without calculus; C. Smith's *Analytical Conics*; Edwards's *Differential Calculus*; Williamson's *Integral Calculus*; Besant's *Hydrostatics*. These were annotated by Macaulay and provided with revision papers at intervals. Beyond this point the order could be varied to suit individual tastes. My sequence, I think, was: Casey's *Sequel to Euclid*; Chrystal's *Algebra II*; Salmon's *Conics*; Hobson's *Trigonometry* (2nd edition, 1897); Routh's *Dynamics of a Particle* (a book of more than 400 pages and containing some remarkably highbrow excursions towards the end); Routh's *Rigid Dynamics*; Spherical Trigonometry (in every possible detail); Murray's *Differential Equations*; Smith's *Solid Geometry*; Burnside and Panton's *Theory of Equations*; Minchin's *Statics* (omitting elasticity, but including attractions, with spherical harmonics, and — of course — an exhaustive treatment of the attractions of ellipsoids).

I had read nearly all of this before the Entrance Scholarship Examination of December 1902. (I was expected to do well, but I found the papers difficult and got only a Minor Scholarship at Trinity.[1] I had had a severe attack of influenza some weeks before, and though I did not feel mentally unfit I certainly must have been.) We were not overtaught and there were no oral lessons, and while anyone *could* go to Macaulay in a difficulty it was on the whole not done. We went up, of course, with paper work at intervals, at first from examples marked by him in the current book, later from our own selections. (There was a weekly problem paper from Wolstenholme's collection, set at one time by him, later by the head boy, who was myself in my last year; if we all failed at a problem it became Macaulay's duty to perform at sight at the blackboard.) The class were encouraged to go to seniors for help, I should say to the great benefit of all concerned. Work directly for the Scholarship Examination was confined to a revision in the term preceding it. (His academic successes, however, were notorious. In his 25 years at the School there were 41 scholarships (34 in Cambridge) and 11 exhibitions; and in the 20 years available these provided four Senior Wranglers, one 2nd, and one 4th. My own period was a peak. G. N. Watson, a year my junior at the school and at Cambridge, was also a Senior Wrangler; incidentally he was fully as precocious as myself. G. R. Blanco-White, a year my senior, was 2nd Wrangler.) Dr. Maxwell Garnett's description of the education as having a University atmosphere is a fair one. Self-reliance being the expected thing we mostly acquired it, and as Macaulay himself did creative work (he became an F.R.S. in 1928) we caught something of the feeling that mathematics was a natural activity.

There was nothing much wrong with my education so far and what was wrong was inherent in the system. Ideally I should have learnt analysis from a French *Cours d'Analyse* instead of from Chrystal and Hobson, but this would have been utterly unconventional. I did not see myself as a pure mathematician (still less as an analyst) until after my

---

[1] I remember being terribly put off in the first paper by sitting opposite a man who was reeling off the questions: I changed my seat for later papers. It must have been Mercer (who was a graduate at Manchester University and was making a fresh start at Cambridge, a not uncommon practice at the time).

I remember also that Cambridge inspired in me an awe equalled by nothing I have felt since.

Tripos Part I, but I had enough instinctive interest in rigour to make me master the chapters of Chrystal on limits and convergence. The work is rigorous (within reasonable limits), and I really did understand, for instance, uniform convergence, but it is appallingly heavy going. (The 2nd edition of Hobson (1897) was a strange mixture, as Macaulay observed in a marginal note, of careful rigour and astonishing howlers[1], but I had done the 'convergence' sections in Chrystal.)

From this point (the Scholarship Examination), however, I wasted my time, except for rare interludes, for $2\frac{1}{2}$ years (8 months at school, 2 academic years at Cambridge). First the 8 months at school. Rightly enough to begin with, I read Smith's *Solid Geometry*: this did not take long, though I recall that while I followed it easily enough I failed to digest it for examination purposes and did very badly in the questions in the final school examination. The best things to do in applied mathematics would have been the 'water, gas and electricity ' subjects. There was probably no suitable text-book on electricity, but Lamb's *Hydrodynamics* was available. The prolonged study of dynamics for some reason (my own fault) stopped short at only the elements of moving axes. In pure mathematics the ideal would again have been more *Cours d'Analyse*. Instead of such things I spent a long time reading Tait's book on the futile subject of Quaternions. Then occurred one of the interludes: I read Harkness and Morley's *Introduction to the Theory of Analytic Functions* (1898). The correct thing to say would be something about the opening up of infinite horizons and a new spirit of approach to mathematics. The cold facts were quite different. I was indeed greatly struck by individual things[2] and a number of them stuck with me for a long time[3]. But no

---

[1] E.g. the fallacious proof that two power series agreeing in value have identical coefficients. On p. 243–4, again, there are remarkable passages. *'If the limit S_n is infinite, or if it is finite but not definite, the series is not convergent.' 'To show that (the general principle of convergence) is sufficient, denote by R_n the infinite series $a_{n+1}+a_{n+2}+\ldots$, the remainder after n terms, then by making r (in $\sum_{n+1}^{n+r}$) infinite, we see that $|R_n| < \epsilon$ if $n \geq m$, hence S has a value between $S_{n+\epsilon}$ (and ϵ is arbitrarily small); also S_n being the sum of a number of finite quantities is finite, hence S is finite. Thus $S_{n+r} - S_n$ can be made as small as we please by making n large enough, therefore $\lim S_n = \lim S_{n+r}$, hence the value of S is definite, being independent of the form of n' (Trivial punctuations to abbreviate; punctuation as in the original.)* Hobson was a professional analyst when he wrote this: it is a case, certainly very extreme, of blind spots and blindly following tradition when writing a text-book — one cannot be reopening questions all the time. I once caught myself in lecture reproducing a very bad test for differentiating under the integral sign, oblivious of the good one I should be using if I were writing a paper.

[2] Much as everyone is struck on first meeting definite integrals by contour integration; this was in fact one of the things.

[3] This led to an incident in my first term at Cambridge. Our lecturer on analysis was rather a martinet. On one occasion I knew exactly what was coming, having read it in H. and M.: I wrote it all down at speed and looked elaborately out of the window.

infinite horizons. I am, as a matter of fact, sceptical about such 'intro-
ductions': they can seem admirable if you know the subject already, but
I don't think they thrill the beginner, at least legitimately; nothing but
the hard technical story is any real good. incidentally the book is a bit
woolly in its account of real numbers; this was perhaps hardly avoidable
in 1898; nothing my generation ever came across (at any rate in English)
had the sharp bracing precision the student gets today.

On coming up to Cambridge (October 1903) I coached for 2 years
(20 months[1]) for Part I of the Tripos with R. A. Herman, contemporary and friend of my father, and the last of the great coaches[2]. The
period is gloomy to look back on. If I am to record new things I acquired which were in any sense worth acquiring, they were moving axes
in dynamics, hydrodynamics, and differential geometry (beyond what
was in Smith). Also small additions to what I knew already in spherical harmonics and complex variable analysis. Electricity was completely
scrappy and I never saw Maxwell's equations[3]. Enthusiasm was touched
just twice, by a stimulating course in the first term by A. N. Whitehead
on the foundations of mechanics, and by an admirable one on differential
geometry given by Herman in his capacity of College lecturer, of which
more later. To be in the running for Senior Wrangler one had to spend
two-thirds of the time practising how to solve difficult problems against
time. I remember that I had then no serious use for lectures, except Herman's; my note-books show that I attended only about half the time,
and in such cases I never looked at the notes again.

It used to be said that the discipline in 'manipulative skill' bore
later fruit in original work. I should deny this almost absolutely — such
skill is very short-winded. My actual experience has been that after a
few years nothing remained to show for it all except the knack, which has
lasted, of throwing off a set of (modern) Tripos questions both suitable
and with the silly little touch of distinction we still feel is called for;
this never bothers me as it does my juniors. (I said 'almost' absolutely;
there could be rare exceptions. If Herman had been put on to some
of the more elusive elementary inequalities at the right moment I can

---

'Are you not taking this down, Sir?' 'I've got it down.' He visibly hesitated whether
to ask to see it (I was not then known to be any good), but in the end said 'I beg
your pardon.' The class thought I had scored, but for myself I felt that he had, by
making me feel a boor.

[1]One long vacation only

[2]The reform of 1910 extinguished almost at once the general practice of
coaching.

[3]It is fair to say that in 2 years I could not use all the available courses (College or
coaching). For completeness I should add that I wasted time on optics and astronomy
(*not* worth knowing) and then practically disregarded them.

imagine his anticipating some of the latest and slickest proofs, perhaps even making new discoveries.)

The old Tripos and its vices are dead; I will not flay them. I do not claim to have suffered high-souled frustration. I took things as they came; the game we were playing came easily to me, and I even felt a satisfaction of a sort in successful craftsmanship.

My detailed career for the 20 months was as follows. I overworked in my first Michaelmas term. On the other hand, I did all but no work in the Lent term (partly because of training for the Lent races); in consequence I took the Trinity March Scholarship examination[1], for which anyhow it was impossible to prepare, feeling at the top of my form, and reversed my failure in the Entrance by coming out top of the list. In June I took 2nd year Mays and came out top (Mercer not sitting). I got full marks in the Analysis paper, my first contact with a startled Hardy, who had just come on the Trinity staff (and was privately coaching Mercer). In my 2nd year the only academic event was the Tripos (which I took while still 19): I was bracketed Senior Wrangler with Mercer.

The Mays Analysis paper reminds me of a fatuous experiment, and I will digress. I lived in Bideford (Devon) and decided to spend part of the Easter vacation buried at Hartland Quay (in superb scenery, and the spot in England most distant from a railway station). The idea was to give up smoking, concentrate on work in the mornings and late afternoons, and 'relax' on poetry and philosophy (Principia Ethica) in the evenings, fortified by strong coffee. (Incidentally my generation worked mainly at night, and 1 o'clock was early to go to bed: there was also a monstrous belief that 8 hours was the minimum a mathematician should work a day; the really virtuous man, by cutting down his sleep, should achieve 10.) My window opened on the sea, which I used as a waste-paper-basket, and on arrival I ceremonially threw my pipes and tobacco into it. Next day I relapsed. The work I got through was very slight, but it consisted in reading the parts of Whittaker's and Watson's *A Course of Modern Analysis* I did not already know, and revising, and this is why analysis was at my fingers' ends in the Mays. The experiment taught me something of the truth that for serious work one does best with a background of familiar routine, and that in the intervals for relaxation one should *be* relaxed. Much could be said on this theme, but this is not the moment for it: I will say, however, that for me the thing to avoid, for doing creative work, is above all Cambridge life, with the constant bright conversation of the clever, the wrong sort of mental stimulus, all

---

[1] For Senior Scholarships, and open to all (including Entrance Scholars) not already Senior Scholars: it is now abolished.

the goods in the front window.

Something about the M.T.I. examination itself. It consisted of 7 papers ('1st 4 days') on comparatively elementary subjects, the riders, however, being quite stiff, followed a week later by another 7 ('2nd 4 days'). A pass on the 1st 4 days qualified for a degree, but the 2nd 4 days carried double the marks, and since it was impossible to revise everything the leading candidates concentrated on the 2nd 4 days[1], in which, moreover, it was generally possible to find enough questions without preparing all subjects. The leaders generally came out pretty level on the 1st 4 days, and the things they used to do I now find almost incredible. It is a lost world, and except for odd accidents I cannot remotely guess what questions I did, but I inherited the mark sheets of my year (1905) from one of the examiners. The marks of the leaders in the 1st 4 days, ignoring the problem paper, were 1350, 1330, 1280, 1230 (the Senior Wranglers 3rd and 4th), followed by 8 more, of whom the last got 990. Full marks were 1930, and the papers were of 10 questions, for the most part with stiff riders. On the problem paper Mercer got 270 out of 760 *for* 18 *questions* (I got only 180). In one paper I got 177 out of 230 for the riders, and I can remember something of this. One whole question was book-work about Carnot's cycle, of which I had not heard. Another was about a condenser, of which I also had not heard, but I reconstructed the question from the answer to the rider. My recollection, however, is that I did *all* that paper apart from Carnot: the marks I dropped must have been for inaccuracy and my notoriously slovenly 'style'[2]. In the 2nd 4 days (ignoring the problem paper) Mercer and I each got about 2050 out of 4500 (each about 330 out of 1340 in the 18 question problem paper). What staggers me most here was a paper (mixed pure and applied) in which I got practically full marks for book-work (290 out of 310, apparently I avoided 'slovenliness' here) *plus* 250 out of 590 for the riders. The marks of the candidates have a frequency graph not at all Gaussian; it is horizontal from the highest point onward. Explanations suggest themselves, but oddly enough the graph of a recent Mays examination is roughly Gaussian.

There is only one other question I am sure of having done, and for the following reason. I began on a question on elementary theory of

---

[1] I did one very bad paper in the 1st 4 days, all optics and astronomy.

[2] I do not take off marks in examinations for slovenliness as such (and always protest against examiners' bleatings that 'the numerical work was slovenly and inaccurate'). Muddled writing in considered work is of course a heinous crime, but at speed and at examination level it is venial. Much nonsense used to be talked about this. It amuses me to recall the man, famous for clear thinking emerging in faultless copperplate even in examinations, and held up to us as a model. In his later career he wrote more bad, muddled, and completely wrong mathematics than anyone before or since.

numbers, in which I felt safe in my school days. It did not come out, nor did it on a later attack. I had occasion to fetch more paper; when passing a desk my eye lit on a heavy mark against the question. The candidate was not one of the leading people, and I half unconsciously inferred that I was making unnecessarily heavy weather; the question then came out fairly easily. The perfectly highminded man would no doubt have abstained from further attack; I wish I had done so, but the offence does not lie very heavily on my conscience.

The M.T. II (1906). This dealt in quite genuine mathematics (and except that Part III, as it is now called, is taken in the 3rd instead of a normal 4th year the examination has been much the same ever since the '80s). I wasted a good deal of time, unluckily in some ways, but partly in the ordinary course of trial and error. Pursuing differential geometry, I embarked on Darboux's *Théorie des Surfaces*, and read 3 of the 4 volumes (i.e. I read 1500 pages). It is a beautiful work, but my initial enthusiasm flagged: it was not my subject. In the examination there were several questions on it; I did them all; but I could have done them the year before, from Herman's lectures. The rest of my studies were in analysis of a sort. One thing Cambridge made almost inevitable for an analyst; intensive study of Legendre functions, and all that. Such 'dictionary' subjects are utterly unsuitable for a good man. It was, however, the more inevitable in that a lecture was provided (by E. W. Hobson; he later wrote a standard text-book): I was the sole member of the class. It amuses me to recall that I could get up this kind of thing completely by heart: several questions were set and I wiped the floor with all of them. I attended, also as sole member of the class, a lecture by E. W. Barnes on his current work on double Γ and ς functions. There was a highly individual lecture by H. F. Baker, on selected points from widely scattered branches of analysis; this was stimulating but not intended to be pedagogic. I presented the more elementary of the two parts constituting 'Elliptic Functions' in the Schedule and 'attended' the course on this by A. Berry. The lectures, however, were at 9 a.m., and I managed to get there only about half the time (working as I did till 2 or 3 in the morning); I never read up the notes, nor did I follow the obvious course of reading a text-book (though we depended far more on text-books and less on lectures than now), and I abandoned the subject for examination purposes. The fact is that I had as yet no sort of idea of what was good for me, and, again, I read no complex function theory proper. Having somehow acquired a working knowledge of Analysis I never read seriously any of the *Cours d'Analyse*. Others have put on record how Jordan first opened their eyes to what real mathematics was; this I missed. But I was also very casual; Picard would certainly have been very good for me. My memory of all this is very hazy and I must have read things I have forgotten. A few weeks before the examination, in the Easter term, I first came across the early volumes of the Borel series, and it was these, in cold fact, that first gave me an authentic thrill: series of positive terms, divergent series, and the volume on integral functions. The first two were irrelevant for the examination; the last I presented officially, but I lost the book, could not conveniently get another copy, and did not prepare it. But now I knew the kind of thing I wanted.

For special reasons I can identify the details about one paper (Friday, June 1, 1906, 9–12), and it interests me, in the light of my later activities, to see what I did *not* then know. There were 6 questions:

1. Elaborate Legendre functions.

2. Multiplication of series.

3. Discontinuous Riemann integrable function.

4. Reversion of power series (with a specific radius of convergence to be established).

5. Conformal representation of an oval on a half-plane.

6. Elliptic functions (ϑ-functions from the part of the subject I was not presenting).

In this the only question I was supposed to know about was number 1 (where of course I got full marks at high speed). By the rules of the game this entitled me to fullest possible marks for the whole paper; but in such cases one was naturally a little nervous about whether the examiners would notice, and this is why I remember everything. I knew something about number 4 unofficially, but not in the precise form given, so my answer was imperfect. About all the rest my ignorance was actual as well as official (though 5 years earlier, at school, I had known number 2).

I was told that I had done very well in the whole examination. I had, however, not yet technically qualified for a degree. In those days a Tripos taken in one's 2nd year did not count at all, and while M.T. II qualified for a degree when taken in one's 4th year it did not do so in one's 3rd. There have always been provisions (Special Graces of the Senate) for dealing with anomalies, however, and I heard about this only by chance.

There were 9 classes, I(1) to III(3). The standard was sometimes preposterous, and the 9 class system went out in 1910 in a blaze of glory. An unusually large and strong field included 6 people afterwards well-known as Professors of mathematics, or an equivalent; class I division 1 was empty. It was empty also the previous year, and had only 1 member the year before that.

To finish with lectures. In my 4th year there were probably few left for me to go to. A. R. Forsyth (Sadleirian Professor) gave a course on differential equations; this did not appeal to me. What I did go to were courses by Whitehead on foundations of geometry and on foundations of mathematics, given for the first time. There were 3 or 4 of us in the class and we found them very exciting. (Whitehead had recently been made a Senior College Lecturer at Trinity, with the duty of giving lectures out of the ordinary run. His stock College lectures, except for the one on the principles of mechanics, were solid and unexciting affairs on applied mathematics: a mathematician can have the duty of being dull — Eddington lecturing on Spherical Trigonometry.) I cannot remember going to any other lectures.

My research began, naturally, in the Long Vacation of my 3rd year, 1906. My director of studies (and tutor) E. W. Barnes suggested the subject of integral functions of order 0. The first idea was to find asymptotic formulae for functions with simple given zeros like $a_n = e^n$; the analytic methods he had been using with success for non-zero order were not working. Incidentally this brought me into touch with another famous and important Borel volume, Lindelöf's *Calcul des Résidus*. There were the best of reasons for the method's not working, as appeared later, but the general suggestion was an excellent one; I rather luckily struck oil at once by switching to more 'elementary' methods, and after that never looked back. The conjecture soon suggested itself that a function of order 0 would, on some large circles, have the property

$$m(r) > (M(r))^{1-r},$$

where $M(r)$ and $m(r)$ are the maximum and minimum moduli. By my elementary methods, at least, this is quite tough, and it took me probably a couple of months. (The corresponding result for non-zero order is that for order less than $\frac{1}{2}$, $m(r)$ is as large as a positive power of $M(r)$ (on some circles). This I could prove only with $\frac{1}{4}$ for $\frac{1}{2}$, and the full result was proved later by A. Wiman by more 'function-theory' methods. A. S. Besicovitch, however, has recently revived the elementary method to prove further extensions.) I sent a longish paper (about functions of order 0) to the London Mathematical Society (Jan. 1, 1907). I should omit a good deal today, but it was not too obscurely written, and the $m > M^{1-\epsilon}$ result is quite respectable. It also contains what I believe is the first instance of a certain 'averaging argument'. (One wants to prove that a function $f(x)$ exceeds a suitable number m at *some* point of a range $0 \le x \le 1$, say. Each individual point is intractable, but the way out sometimes exists that the average of $f(x)$ over $(0, 1)$ can be shown to exceed an m; then *some* x, though unidentifiable, must have $f(x) > m$.) The referees disagreed, one being violently unfavourable (by the time I learned in later life who he was I had disinterestedly come to think him a bit of an ass). Hardy was appointed as third referee and the paper was duly published. I have not since had trouble with papers, with the single exception that the Cambridge Philosophical Society once rejected (quite wrongly) one written in collaboration with Hardy.

Barnes was now encouraged to suggest a new problem: 'Prove the Riemann Hypothesis'. As a matter of fact this heroic suggestion was not without result; but I must begin by sketching the background of $\varsigma(s)$ and prime numbers in 1907, especially so far as I was myself concerned. I had met $\varsigma(s)$ in Lindelöf, but there is nothing there about primes, nor had I the faintest idea there was any connexion; for me the R.H. was famous, but only as a problem in integral functions; and all this took place in the Long Vacation when I had no access to literature, had I suspected there was any. (As for people better instructed, only some had heard of Hadamard's paper, and fewer still knew of de la Vallée Poussin's in a Belgian journal. In any case, the work was considered very sophisticated and outside the main stream of mathematics. The famous paper of Riemann is included in his collected works; this states the R.H., and the extraordinary, but unproved, 'explicit formula' for $\pi(x)$; the 'Prime Number Theorem' is not mentioned, though it is doubtless an easy guess granted the explicit formula. As for Hardy in particular, he told me later that he 'knew' the P.N.T. had been proved, but he thought by Riemann. All this was transformed at a stroke by the appearance of Landau's book in 1909.)

I remembered the Euler formula $\sum n^{-s} = \Pi(1 - p^{-s})^{-1}$; it was introduced to us at school, as a *joke* (rightly enough, and in excellent taste). (Oddly enough it is not in Chrystal's *Algebra*; but in the 'convergence' chapter there is an example, with references: $\sum f(p)$ is convergent if $\sum f(n)/\log n$ is. The n's are however misprinted as p's. Against the resulting false statement I find a note made by me in 1902, query $f(p) = l/p$: I was sure in 1902 that $\sum 1/(p \log p)$ converges — it is actually not too big a jump from the Euler product.) In the light of Euler's formula it is natural to study $P(s) = \sum p^{-s}$. I soon saw that if the P.N.T. were true 'with error about $\sqrt{x}$' the R.H. would follow. Now at that time, and for anyone unacquainted with the literature, there was no reason to expect any devilment in the primes. And the $\sqrt{x}$ seems entirely natural, for the reason that a proper factor of n cannot exceed $\sqrt{n}$. So I started off in great excitement and confidence, and only after a week or so of agony came to realize the true state of things. There was, however, a consolation prize. It occurred to me to try the reverse argument: I assumed R.H., operated (in the line of least resistance) with the integral function

$$H\left\{\left(1+\frac{2}{p}\right)e^{-z/p}\right\}$$

and successfully deduced the P.N.T. (This was just in time for my first Fellowship dissertation (September 1907); I suppressed it the following year.)

I have a clear recollection of my youthful views about the P.N.T., and they illustrate the uncertainty of judgment and taste in a beginner in a field with no familiar land-marks. I was thrilled *myself*; but didn't feel at all sure how the result would appeal to others, and if someone had said, 'not bad, but of course very special, not "proper" mathematics', I should have meekly acquiesced. Hardy (a junior Fellowship Elector at the time) told me much later that he had 'courageously' said at the time that it was the best thing in the dissertation, though without realizing it was submitted as original. The dissertation as a whole was well received, and though I was passed over for a man at his last shot there was a gentleman's agreement that I should be elected next time.

From October 1907 to June 1910 I was Richardson lecturer at Manchester University. At £250 this was better than the usual £150 or £120, and I was advised to take it, but it was a great mistake. I could have stayed in Cambridge as a Research Scholar, and was soon offered the Allen Scholarship (incidentally tenable with a Fellowship if one got that later), but refused it to stay at Manchester. I did not gain financially, but felt I needed a change from Cambridge. If an austere desire for working at full stretch was also a motive it was fulfilled. My work was as follows. 3 hours lecturing to a 'failed Matriculation' class (the University earned fees by this); 3 hours to superior Intermediate; 3 (possibly 2) hours to pupil teachers on 'principles of Mathematics' (an 'Education' stunt, naturally a complete failure), 2 hours class-work with 3rd year Honours class, and 3 hours full-dress lectures to them. Most of the unoccupied time during the mornings was spent in a sort of 'class-work': 12 to 20 students sat doing examples, to be helped out when they got stuck; it is an admirable system (for the students). Beyond this there was much paper work from the large elementary classes. In any case, and whatever the details, what happened was 4 hours work of one sort or another on M., W., F. mornings, 3 hours Tu., Th. mornings; after lunch paper work and some lecture preparation done in a private room at the University and lasting from 2.30 to 4.0 or 4.30. (Elementary lectures we learned, of course, to deliver with a minimum of preparation, on occasion extempore.) Saturdays were free. But while for most of the staff the day's work ended by 4.30, I had high pressure work on top of the low pressure mountain. The 3rd year Honours class at that time got what was in spirit the most liberal mathematical education in the country. Unhampered by the official examinations, which were made to yield the results known to be right, we aimed at doing a few selected things really properly, dealt with some utilitarian stuff in class-work, and did not try to cover everything. The Pure side of this was my responsibility, and I had a completely free hand. One of my selections was Differential Geometry. This gave comparatively little trouble, since I stuck slavishly to my notes of Herman, except for a necessary dilution. (Many years later I mentioned this to Hardy, who confessed in return that when he found being Professor of Geometry at Oxford involved giving actual lectures in geometry he did exactly the same thing.) For the rest my lectures were in analysis. These called for as much preparation as any I have given since, and I had to prepare them in the evenings. Hardy's *Pure Mathematics* and Bromwich's *Infinite Series* were not available the first year. I must have found Jordan no use for what I wanted; Goursat's *Cours* I could have used to some extent, were it not for the almost incredible fact that I was unaware of its existence. It is hard to realize now the difficulty of planning a logical order that would not unexpectedly let one down (and my admiration for Bromwich's performance was unbounded). I aimed only at teaching a working efficiency (no elegance, but full rigour — and we dealt even in repeated infinite integrals), but it was exceedingly hard going. The lectures were fairly successful, and temporarily seduced Sydney Chapman into becoming an analyst. (I added to my difficulties by being one of the most feckless young men I know of; my lecture notes were not unnaturally scrawled, but they were on odd sheets and too chaotic to be used another year.) It remains to add to this story that the two long terms were 10-week, the only remission being that the Long Vacation began reasonably early in June. Work at this pressure (apart from my special difficulties) was the accepted thing, and research was supposed to be done in one's leisure: I remember one Easter vacation, when I was worn out and could not force myself to work, suffering pangs of conscience over my laziness. 'Young men of today don't know what work is.' I should add that H. Lamb (doubling the parts of Pure and Applied Professors) did his full share of the work, and showed me many kindnesses.

I joined the Trinity staff in October 1910 (succeeding Whitehead). This coincided with new mathematical interests. Landau's book on analytical number theory made exciting reading, and stimulated me to some ideas on the ς-function, but I need not say anything about this. I have, however, some vivid, and to me amusing, recollections of the discovery

The Extension of Tauber's Theorem.

By J. E. Littlewood

Received 28/IX/10
Read 10/XI/10

§1. Tauber's theorem asserts that the existence of $\lim_{x \to 1} \Sigma a_n x^n$, together with $na_n \to 0$, implies the convergence of Σa_n. The main object of the present paper is to show that the theorem is valid under the wider condition $|na_n| < K$, where K is any positive constant; and, on the other hand, that no further extension in this direction is possible.

It is possible, without any very serious modification in the line of argument, to establish similar conclusions with the more general series $\Sigma a_n x^{n^k}$ in the place of $\Sigma a_n x^n$. We shall show, in fact, that

$$\Sigma a_n x^{n^k} \to S, \quad (k \text{ any positive constant}), \text{ and } |na_n| < K,$$

together imply the convergence of Σa_n; and that, whatever k may be, and whatever function $\phi(n)$, tending to ∞ with n, may be chosen,

$$\Sigma a_n x^{n^k} \to s \quad \text{and} \quad |na_n| < \phi(n)$$

A passage from the ms. of the 'Abel-Tauber paper', published later in a thoroughly revised form under the title *The converse of Abel's theorem on power series*.

of the proof of the 'Abel-Tauber Theorem' ('If $\lim_{x \to 1} \sum a_n x^n = s$ and $a_n = O(1/n)$ then $\sum a_n$, converges to s'). This happened at Bideford in the Easter vacation of 1911. The problem had quite certainly been suggested by Hardy, but I was unaware that he had proved the (weaker) 'Cesáro-Tauber'. This is very strange. In the first place he had told me about it; but I suppose at a time when I had not begun to think actively in that field. On the other hand, I had at that time in high degree the flair of the young for tracking down any previous experience that might bear on the problem in hand; this must have been out of action. But however strange, it was providential — *wen Gott betrügt ist wohl betrogen*. The main theorem depends on two separate ideas, and one of them is the connexion between 3 (or more) successive derivates (if $f = o(1)$ and $f'' = O(1)$ then $f' = o(1)$[1].). I began on the Cesáro-Tauber and in the course of finding a proof was led to the derivates theorem: but for this the derivates theorem would never have emerged out of the rut of the established proof (which differed a good deal), and without it I should never have got the main theorem. (The derivates theorem was actually known, but buried in a paper by Hadamard on waves.) It is of course good policy, and I have often practised it, to begin without going too much into the existing literature.

The derivates theorem enables one to reject certain parts of the thing one wants to tend to zero. One day I was playing round with this, and a ghost of an idea entered my mind of making r, the number of differentiations, *large*. At that moment the spring cleaning that was in progress reached the room I was working in, and there was nothing for it but to go walking for 2 hours, in pouring rain. The problem seethed violently in my mind: the material was disordered and cluttered up with irrelevant complications cleared away in the final version, and the 'idea' was vague and elusive. Finally I stopped, in the rain, gazing blankly for minutes on end over a little bridge into a stream (near Kenwith wood), and presently a flooding certainty came into my mind that the thing was done. The 40 minutes before I got back and could verify were none the less tense.

On looking back this time seems to me to mark my arrival at a reasonably assured judgment and taste, the end of my 'education'. I soon began my 35-year collaboration with Hardy.

---

[1]See p. 54, and footnote 2

REVIEW OF RAMANUJAN'S COLLECTED PAPERS[1]

Collected Papers of Srinivasa Ramanujan. Edited by G. H. Hardy, P. V. Seshu Aiyar, and B. M. Wilson. Pp. *xxxvi* + 355. 30s. net. 1927. (Cambridge Univ. Press.)

Ramanujan was born in India in December 1887, came to Trinity College, Cambridge, in April 1914, was ill from May 1917 onwards, returned to India in February 1919, and died in April 1920. He was a Fellow of Trinity and a Fellow of the Royal Society.

Ramanujan had no university education, and worked unaided in India until he was twenty-seven. When he was sixteen he came by chance on a copy of Carr's *Synopsis of Mathematics*; and this book, now sure of an immortality its author can hardly have dreamt of, woke him quite suddenly to full activity. A study of its contents is indispensable to any considered judgement. It gives a very full account of the purely formal side of the integral calculus, containing, for example, Parseval's formula, Fourier's repeated integral and other 'inversion formulae', and a number of formulae of the type recognizable by the expert under the general description '$f(\alpha) = f(\beta)$ if $\alpha\beta = \pi^2$'. There is also a section on the transformation of power series into continued fractions. Ramanujan somehow acquired also an effectively complete knowledge of the formal side of the theory of elliptic functions (not in Carr). The matter is obscure, but this, together with what is to be found in, say, Chrystal's *Algebra*, seems to have been his complete equipment in analysis and theory of numbers. It is at least certain that he knew nothing of existing methods of working with divergent series, nothing of quadratic residues, nothing of work on the distribution of primes (he may have known Euler's formula $\Pi(1 - p^{-s})^{-1} = \sum n^{-s}$, but not any account of the ς-function). Above all, he was totally ignorant of Cauchy's theorem and complex function-theory. (This may seem difficult to reconcile with his complete

---

knowledge of elliptic functions. A sufficient, and I think a necessary, explanation would be that Greenhill's very odd and individual *Elliptic Functions* was his text-book.)

The work he published during his Indian period did not represent his best ideas, which he was probably unable to expound to the satisfaction of editors. At the beginning of 1914, however, a letter from Ramanujan to Mr. Hardy (then at Trinity, Cambridge) gave unmistakeable evidence of his powers, and he was brought to Trinity, where he had three years of health and activity. (Some characteristic work, however, belongs to his two years of illness.)

I do not intend to discuss here in detail the work for which Ramanujan was solely responsible (a very interesting estimate is given by Prof. Hardy, p. *xxxiv*). If we leave out of account for the moment a famous paper written in collaboration with Hardy, his definite contributions to mathematics, substantial and original as they are, must, I think, take second place in general interest to the romance of his life and mathematical career, his unusual psychology, and above all to the fascinating problem of how great a mathematician he might have become in more fortunate circumstances. In saying this, of course, I am adopting the highest possible standard, but no other is appropriate.

Ramanujan's great gift is a 'formal' one; he dealt in 'formulae'. To be quite clear what is meant, I give two examples (the second is at random, the first is one of supreme beauty):

$$p(4) + p(9)x + p(14)x^2 + \ldots = 5\frac{\{(1 - x^5)(1 - x^{10})(1 - x^{15})\ldots\}^5}{\{(1 - x)(1 - x^2)(1 - x^3)\ldots\}^6},$$

where $p(n)$ is the number of partitions of n;

$$\int_0^\infty \frac{\cos \pi x}{\{\Gamma(\alpha + x)\Gamma(\alpha - x)\}^2}dx = \frac{1}{4\Gamma(2\alpha - 1)\{\Gamma(\alpha)\}^2} \quad (\alpha > \tfrac{1}{2}).$$

But the great day of formulae seems to be over. No one, if we are again to take the highest standpoint, seems able to discover a radically new type, though Ramanujan comes near it in his work on partition series; it is futile to multiply examples in the spheres of Cauchy's theorem and elliptic function theory, and some general theory dominates, if in a less degree, every other field. A hundred years or so ago his powers would have had ample scope. Discoveries alter the general mathematical atmosphere and have very remote effects, and we are not prone to attach great weight to rediscoveries, however independent they seem. How much are we to allow for this, how great a mathematician might Ramanujan have been 100 or 150 years ago; what would have happened if he had come into touch with Euler at the right moment? How much does lack of education matter? Was it formulae or nothing, or did he develop in the direction he did only because of Carr's book — after all, he learned later to do new things well, and at an age mature for an Indian? Such are the questions Ramanujan raises; and everyone has now the material to judge them. The letters and the lists of results announced without proof are the most valuable evidence available in the present volume; they suggest, indeed, that the note-books would give an even more definite picture of the essential Ramanujan, and it is very much to be hoped that the editors' project of publishing them *in extenso* will eventually be carried out.

Carr's book quite plainly gave Ramanujan both a general direction and the germs of many of his most elaborate developments. But even with these partly derivative results one is impressed by his extraordinary profusion, variety, and power. There is hardly a field of formulae, except that of classical number-theory, that he has not enriched, and in which he has not revealed unsuspected possibilities. The beauty and singularity of his results is entirely uncanny. Are they odder than one would expect things selected for oddity to be? The moral seems to be that we never expect enough; the reader at any rate experiences perpetual shocks of delighted surprise. And if he will sit down to an unproved result taken at random, he will find, if he can prove it at all, that there is at lowest some 'point', some odd or unexpected twist. Prof. Watson and Mr. Preece have begun the heroic task of working through the unproved statements; some of their solutions have appeared recently in the *Journal of the London Mathematical Society*, and these strongly encourage the opinion that a complete analysis of the note-books will prove very well worth while.

There can, however, be little doubt that the results showing the most striking originality and the deepest insight are those on the distribution of primes (see pp. *xxii–xxv, xxvii*, 351, 352). The problems here are not in origin formal at all; they concern approximate formulae for such things as the number of primes, or of integers expressible as a sum of two squares, less than a large number x; and the determination of the orders of the errors is a major part of the theory. The subject has a subtle function-theory side; it was inevitable that Ramanujan should fail here, and that his methods should lead him astray; he predicts the approximate formulae, but is quite wrong about the orders of the errors. These problems tax the last resources of analysis, took over a hundred years to solve, and were not solved at all before 1890; Ramanujan could not possibly have achieved complete success. What he did was to perceive that an attack on the problems could at least be begun on the formal side, and to reach a point at which the main results become plausible. The formulae do not in the least lie on the surface, and his achievement, taken as a whole, is most extraordinary.

If Carr's book gave him direction, it had at least nothing to do with his *methods*, the most important of which were completely original. His intuition worked in analogies, sometimes remote, and to an astonishing extent by empirical induction from particular numerical cases. Being without Cauchy's theorem, he naturally dealt much in transformations and inversions of order of double integrals. But his most important weapon seems to have been a highly elaborate technique of transformation by means of divergent series and integrals. (Though methods of this kind are of course known, it seems certain that his discovery was quite independent.) He had no strict logical justification for his operations. He was not interested in rigour, which for that matter is not of first-rate importance in analysis beyond the undergraduate stage, and can be supplied, given a real idea, by any competent professional. The clear-cut idea of what is *meant* by a proof, nowadays so familiar as to be taken for granted, he perhaps did not possess at all. If a significant piece of reasoning occurred somewhere, and the total mixture of evidence and intuition gave him certainty, he looked no further. It is a minor indication of his quality that he can never have *missed* Cauchy's theorem. With it he could have arrived more rapidly and conveniently at certain of his results, but his own methods enabled him to survey the field with an equal comprehensiveness and as sure a grasp.

I must say something finally of the paper on partitions (pp. 276-309) written jointly with Hardy. The number $p(n)$ of the partitions of n increases rapidly with n, thus:

$$p(200) = 3972999029388.$$

The authors show that $p(n)$ is the integer nearest

(1)
$$\frac{1}{2\sqrt{2}} \sum_{q=1}^{\nu} \sqrt{q} A_q(n) \psi_q(n),$$

where $A_q(n) = \sum \omega_{p,q} e^{-2np\pi i/q}$, the sum being over p's prime to q and less than it, $\omega_{p,q}$ is a certain $24q$-th root of unity, ν is of the order of $\sqrt{n}$, and

$$\psi_q(n) = \frac{d}{dn}\left(\exp\{C\sqrt{(n - \tfrac{1}{24})}/q\}\right), \quad C = \pi\sqrt{\frac{2}{3}}.$$

We may take $\nu = 4$ when $n = 100$. For $n = 200$ we may take $\nu = 5$; five terms of the series (1) predict the correct value of $p(200)$. We may always take $\nu = \alpha\sqrt{n}$ (or rather its integral part), where α is any positive constant we please, provided n exceeds a value $n_0(\alpha)$ depending only on α.

The reader does not need to be told that this is a very astonishing theorem, and he will really believe that the methods by which it was established involve a new and important principle, which has been found very fruitful in other fields. The story of the theorem is a romantic one. (To do it justice I must infringe a little the rules about collaboration. I therefore add that Prof. Hardy confirms and permits my statements of bare fact.)   One of Ramanujan's Indian

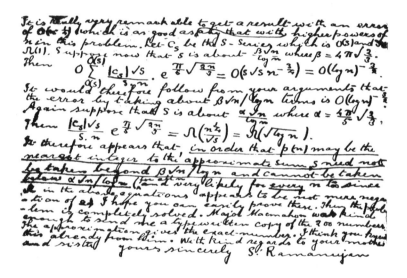

A postcard from Ramanujan to Hardy in December 1916 illustrating some of Littlewood's comments.

conjectures was that the first term of (1) was a very good approximation to $p(n)$; this was established without great difficulty. At this stage the $n - \frac{1}{24}$ was represented by a plain n — the distinction is irrelevant. From this point the real attack begins. The next step in development, not a very great one, was to treat (1) as an 'asymptotic' series, of which a fixed number of terms (e.g. $\nu = 4$) were to be taken, the error being of the order of the next term. But from now to the very end Ramanujan always insisted that much more was true than had been established: 'There must be a formula with error $O(1)$.' This was his most important contribution; it was both absolutely essential and most extraordinary. A severe numerical test was now made, which elicited the astonishing facts about $p(100)$ and $p(200)$. Then ν was made a function of n; this *was* a very great step, and involved new and deep function-theory methods that Ramanujan obviously could not have discovered by himself. The complete theorem thus emerged. But the solution of the final difficulty was probably impossible without one more contribution from Ramanujan, this time a perfectly characteristic one. As if its analytical difficulties were not enough, the theorem was entrenched also behind almost impregnable defences of a purely formal kind. The form of the function $\psi_q(n)$ is a kind of indivisible unit; among many asymptotically equivalent forms it is essential to select exactly the right one. Unless this is done at the outset, and the $-\frac{1}{24}$ (to say nothing of the d/dn) is an extraordinary stroke of formal genius, the complete result can never come into the picture at all. There is, indeed, a touch of real mystery. If only we *knew* there was a formula with error $O(1)$, we might be forced, by slow stages, to the correct form of ψ_q. But why was Ramanujan so certain there *was* one? *Theoretical* insight, to be the explanation, had to be of an order hardly to be credited. Yet it is hard to see what numerical instances could have been available to suggest so strong a result. And unless the form of ψ_q was known already, *no* numerical evidence could suggest anything of the kind — there seems no escape, at least, from the conclusion that the discovery of the correct form has a single stroke of insight. We owe the theorem to a singularly happy collaboration of two men, of quite unlike gifts, in which each contributed the best, most characteristic, and most fortunate work that was in him. Ramanujan's genius did have this one opportunity worthy of it.

The volume contains a biography by the second of the editors, and the obituary notice by Prof. Hardy. These give quite a vivid picture of Ramanujan's interesting and attractive personality. The mathematical editors have done their work most admirably. It is very unobtrusive, the reader is told what he wants to know at exactly the right moment, and more thought and bibliographical research must have gone into it than he is likely to suspect.

LARGE NUMBERS[1]

§1. The problem how to express very large numbers is discussed in 'The Sand-Reckoner' of Archimedes. Grains of sand being proverbially 'innumerable', Archimedes develops a scheme, the equivalent of a 10^n notation in which the 'Universe', a sphere reaching to the sun and calculated to have a diameter less than 10^{10} stadia, would contain, if filled with sand, fewer grains than '1000 units of the seventh order of numbers', which is 10^{51}. [A myriad-myriad is 10^8; this is taken as the base of what we should call exponents, and Archimedes contemplates 10^8 'periods', each containing 10^8 'orders' of numbers; the final number in the scheme is $10^{8 \cdot 10^{15}}$.] The problem of expression is bound up with the invention of a suitable *notation*; Archimedes does not have our a^b, with its potential extension to $a^{a^{a^{\cdot^{\cdot^{\cdot}}}}}$. We return to this question at the end; the subject is not exhausted.

§2. Certain ancient Indian writings reveal an awestruck obsession with ideas of immense stretches of time. See Buckle's *History of Civilisation in England*, pp. 121–124 (2nd edition). (I *thought* the following came from there; I cannot have invented it, surely.)

There is a stone, a cubic mile in size, a million times harder than diamond. Every million years a very holy man visits it to give it the lightest possible touch. The stone is in the end worn away. This works out at something like 10^{35} years; poor value for so much trouble, and an instance of the 'debunking' of popular immensities.

§3. The Greeks made an enormous leap of the imagination in conceiving the heavenly bodies as objects dispersed in 'ordinary' space. A similar if lesser leap was needed to initiate the now familiar geological

---

[1]Reprinted from the *Mathematical Gazette*, July 1948, Vol. *XXXII*, No. 300. Additions in square brackets.

arguments about erosion and the like. It is easy to imagine Archimedes doing this, but so far as I know no Greek did. It can be a mildly entertaining exercise to check, for example, the scooping out of a valley by a trickle of stream, unthinkable to common sense. A twentieth of an inch a year is a mile in 10^6 years; this, if continued, would be a thousand miles in 10^9 years. (These times are natural units; the second is comparable with the age of the earth, the first is the time it takes to turn an ape into a Ph.D.)

Newton estimated the distance of Sirius (in astronomical units), assuming it to be comparable with the sun. His method was to compare Sirius and Saturn, guessing (correctly) the albedo of the latter.

§4. The next two items complete my references to the past. The first is the accuracy of Tycho Brahe's angular measurements. These were correct to 1', which I find surprising (Hipparchus's to 4'). The other is the Samos tunnel, described in Farrington's very interesting Pelican book, *Greek Science*, p. 37. Made at about the time of Pythagoras, it was 900 yards long and was begun at opposite ends; the junction in the middle is shown by modern digging to have been within a couple of feet. I am not concerned with the mild ideological axe-grinding of the book, but it is surely simplest to disbelieve that this was an achievement of surveying technique. The relevant *principles* of similar triangles existed since Thales, but I find the instrumental accuracy incredible. I can, on the other hand, easily believe in a line of posts over the hill, or at a pinch in sighting on a star from opposite sides.

§5. I come to modern times, but continue the topic of measurements. We all know that measured parallaxes deal in quantities of the order of $0''.001$ (average error $0''.025$); does every reader realize that this is the angle subtended by a penny 4000 miles away?

Astronomical measurement of time provides the greatest number of significant figures used in scientific calculation; measurement is to $.001''$, it is possible to deal in 10^2 or 10^3 years, and with a couple of extra figures for purposes of calculation we have a total of 15.

I once asked Eddington what accuracy was possible in measuring the angular separation of widely separated stars. To the outsider the mechanical difficulties seem enormous, as do those of dealing with refraction; the answer (given instantly) was $0''.1$, which I for one found very surprising.

I will hand on another surprise. The principles of an Ordnance Survey clearly involve, first, something of the nature of a 'rod', which is placed in successive positions until we arrive, secondly, at a 'base', from which we carry on by angular measurements. The questions are: what are the most efficient lengths of rod and base? The 'rod' is a metal ribbon 130 inches long, which is much what anyone would expect; but the length of the base, which is 9 miles, is about 100 times what I should have guessed. In such matters, of course, the really determining difficulty is apt to be something not obvious and interesting, but unexpected and dull: apparently the trouble is that of placing a theodolite accurately over the right spot.

§6. We shall presently be considering multiple exponentials and we must consider the principles of their 'approximate' nature. Since 'order' has a technical meaning not suited to our purposes, we will speak of 'types' of numbers,

$$N_1 = 10^{10}, \quad N_2 = 10^{10^{10}}, \ldots, \quad N_n = 10^{N_{n-1}},$$

which we describe as of type $1, 2, \ldots, n, \ldots$ We further describe

$$10^{10^{10^{4 \cdot 7}}},$$

for example, as of type $2 \cdot 47$, and write it $N_{2.47}$. This makes the point that its type is between 2 and 3; there is a slight inaccuracy in that it is $N_{2.0}$ and not $N_{2.1}$ that is N_2 — we may ignore this. We also call it $N_2(4 \cdot 7)$ when we wish to express its mathematical form in brief notation: note that N_n is $N_n(1)$. The number 10^{79}, which (with apologies for the small letter) we will call u, as being the number of ultimate particles in the Universe, is $N_{1.19}$.

The principle I now wish to establish is sufficiently illustrated by the following instances. A number of type 2 or over is 'practically unaltered' by being squared; a number of type 3 or over is what we may fairly call 'unaltered' by being raised even to the power u. In fact, while

$$N_2 = 10^{10^{10}},$$

we have

$$N_2^2 = 10^{10^{10.3}};$$

and while

$$N_3 = N_3(1),$$

we have

$$N_3^u = N_3(1 + 7 \cdot 9 . 10^{-9}).$$

Again, N_2 is hardly altered by having its bottom 10 changed into u, and is 'unaltered' by having it changed into 2. Another constantly relevant point is that for an N of type 1 or over there is 'no difference' between $N!$ (or N^N) and 2^N.

We may sum up these considerations as the 'principle of crudity': the practical upshot is that in estimating a number $a^{b^{c^{\cdot^{\cdot^{\cdot}}}}}$ it is worth taking trouble to whittle down the top index, but we can be as crude as we like about things that bear only on the lowest ones.

§7. I come now to numbers directly connected with daily life (what I mean by 'indirectly' will appear in §11). The range from just perceptible to just tolerable sound (at the same pitch, and where sensitivity is maximal) is over 10^{12}. In the case of light the range is (as we should expect) even greater. The surface of the sun has $6 . 10^5$ times the brightness of the full moon (incidentally the sun is $5 . 10^6$ times as bright as the *half* moon). A sandy surface lit by the full moon is accordingly in a similar relation with the surface of the full moon. Anyone who has walked on a country road on a moonless night with heavy cloud knows that one can still perceive the road or objects on it (I am not myself satisfied that anyone has properly explained where the light comes from); there must be a new factor of at least 10^3 (I should say 10^4 or more), the total being 10^{14} or 10^{15}.

At one time it was possible to buy 10^{13} ergs for 4d; nothing about energy of mass, merely the British Heat Unit: the erg is, of course, absurdly small, and the mechanical equivalent of heat very large.

Coincidences and Improbabilities

§8. Improbabilities are apt to be overestimated. It is true that I should have been surprised in the past to learn that Professor Hardy had joined the Oxford Group. But one could not say the adverse chance was 10^6 : 1. Mathematics is a dangerous profession; an appreciable proportion of us go mad, and then this particular event would be quite likely.

A popular newspaper noted during the 1947 cricket season that two batsmen had each scored 1111 runs for an average of $44 \cdot 44$. Since it compared this with the monkeys' typing of *Hamlet* (somewhat to the disadvantage of the latter) the event is worth debunking as an example of a common class (the same paper later gave a number of similar cases). We have, of course, to estimate the probability of the event happening at some time during the season. Take the 30 leading batsmen and select a pair A, B of them. At some moment A will have played 25 complete innings. The chance against his score then being 1111 is say 700 : 1. The chance against B's having at that moment played 25 innings is say 10 : 1, and the further chance that his score is 1111 is again 700 : 1. There are, however, about 30.15 pairs[1]; the total adverse chance is $10 . 700^2/(30.15)$, or about 10^4 : 1. A modest degree of surprise is legitimate.

A report of holding 13 of a suit at Bridge used to be an annual event. The chance of this in a given deal is $2 \cdot 4 . 10^{-9}$; if we suppose that $2 . 10^6$ people in England each play an average of 30 hands a week the probability is of the right order. I confess that I used to suppose that Bridge hands were not random, on account of inadequate shuffling; Borel's book on Bridge, however, shows that since the distribution within the separate hands is irrelevant the usual procedure of shuffling is adequate. (There is a marked difference where games of Patience are concerned: to destroy all organisation far more shuffling is necessary than one would naturally suppose; I learned this from experience during a period of addiction, and have since compared notes with others.)

I sometimes ask the question: what is the most remarkable coincidence you have experienced, and is it, for *the* most remarkable one, remarkable? (With a lifetime to choose from, 10^6 : 1 is a mere trifle.) This is, of course, a subject made for bores, but I own two, one starting at the moment but debunkable, the other genuinely remarkable. In the latter a girl was walking along Walton St. (London) to visit her sister, Florence Rose Dalton, in service at number 42. She passed number 40 and arrived at 42, where a Florence Rose Dalton was cook (but absent for a fortnight's holiday, deputised for by her sister). But the house was 42 Ovington Sq. (the exit of the Square narrows to road width), 42 Walton St. being the house next further on. (I was staying at the Ovington Sq. house and heard of the occurrence the same evening.) In the other, 7 ships in Weymouth Harbour at the beginning of a 3 mile walk had become 6 when we sat down to rest: the 6 were riding parallel at their anchors, but the two-masted 7th had aligned itself exactly behind a mast of one of the 6. A shift of 5 yards clearly separated the masts. The chance against stopping in the right 10 yards is 600 : 1; that against the ship being end on about 60 : 1; in all about $4 . 10^4$: 1; the event is thus comparable to the cricket average both in striking impact and real insignificance.

There must exist a collection of well-authenticated coincidences, and I regret that I am not better acquainted with them. Dorothy Sayers in

---

[1] Note that it is pairs and not ordered pairs that are relevant

Unpopular Opinions, cites the case of two negroes, each named Will West, confined simultaneously in Leavenworth Penitentiary, U.S.A. (in 1903), and with the same Bertillon measurements. (Is this really credible?)

Eddington once told me that information about a new (newly visible, not necessarily unknown) comet was received by an Observatory in misprinted form; they looked at the place indicated (no doubt sweeping a square degree or so), and saw a new comet. (Entertaining and striking as this is the adverse chance can hardly be put at more than a few times 10^6.)

§9. We all remember the schoolboy doodle of tracing a pencil line down a printed page through the spaces between words. Suppose we take a small-print encyclopedia with about 100 lines to the page, and slash a line through at random. With a 5 : 1 chance against succeeding at a given line the chance against performing this doodle is 10^{70}. My next instance is perhaps off the main track. There is a certain procedure by which a conjuror may perform the apparently impossible. A card, say the Ace of Spades, being selected the conjuror places the pack on the table and asks the subject to think of a number less than 100. There is a very fair chance that he will select 37; in this case he is told to count down and take the 37th card (which is the Ace): if another number is selected *the conjuror does some other trick.* (A milder form deals with numbers less than 10; the selection is very likely to be 7, and if not, then 3; with 9 cards, and the Ace 7th he succeeds outright in the first case and can proffer the inverted pack in the second.)

In my present category belongs the chance typing of *Hamlet* by the monkeys. With say 27,000 letters and spaces to be typed and say 35 keys the adverse chance is $35^{27000} < N_{1.5}$.

Games

§10. Suppose that in a game of position there are p possible positions $P_1, P_2, \ldots, P_p$. A game is a finite sequence of P's, each derived from the preceding by a 'move' in accordance with the rules. p is generally of type slightly greater than 1, and the number of games may consequently be comparable with $p!$ or 2^p, which brings us for the first time to a type above 2. The crudity principle will be in operation.

In Chess, a game is a draw if the same position occurs for the 3rd time in all. (As a matter of fact the game continues unless one of the players exercises the right to claim a draw; to avoid the consequent infinity we will suppose there *is* a draw.) The rules do not say whether for this purpose the men ('man' = 'piece or pawn') retain their identity; we shall suppose that they do.[1]

What is the chance that a person A, ignorant of the rules will defeat C, the world champion? Suppose that in practice C, in 1 out of n of his games, loses in not more than m moves. We suppose that A knows that when it is his turn to move he must place one of his men on an unoccupied square, or on an enemy-occupied square, with capture. When he has $N \leq 16$ men he has a choice of $N^{64-N} \leq M = 16^{48}$ actions. There is, in effect, once in n games, a sequence of m actions leading to victory; his chance in all is better than $1 : nM^m$. The number n hardly matters if we can reduce m; if we may suppose[2] that m is 20 for $n = 10^6$, A has a chance better than $10^{-122} = 1/N_{1\cdot21}$ (more likely than two in succession of the doodles of §9).

What is the number of possible games of Chess? It is easy to give an upper bound. A placing (legal or not) of men on the squares of the board we will call an 'arrangement', A, one possible in Chess we will call a 'position', P. A change from one A to another we call a 'shift', and a legal Chess move (from a P) a 'move'. With N men in all on the board there are (with the 'individuality' convention) $64!/(64 - N)!$ A's. (As a matter of fact, since *all* pawns can be promoted, it is possible for something like this number to be actually P's (for some N); the main legal bars are that the K's must not be contiguous, nor both in check, and that if there are 10 white (or black) bishops, they cannot all be on squares of one colour). The number of sets of men (irrespective of their placing) composed of pawns and pawns promoted (to Q, R, B or Kt) is

$$5^{16} + 5^{15} + \ldots + 5^1 < 2.10^{11}.$$

Since the two K's must be present the number of sets of *pieces* other than those promoted from pawns is

$$_{14}C_1 + _{14}C_2 + \ldots + _{14}C_{14} < 14._{14}C_7 < 5.10^4;$$

hence the number of sets of N men is for every N less than 10^{16}.

The number of moves possible from a P (or for that matter an A) is at most

$$\mu = 9.28 + 2.14 + 2.14 + 2.8 + 8 = 332.$$

---

[1] I learn from Mr. H. A. Webb that in one of Blackburne's games a position recurred for a second time, but with a pair of rooks interchanged. Each player expected to win, otherwise (as Blackburne said) a delicate point for decision would have arisen.

[2] We may suppose that C (in the light of A's performance!) does not suspect the position, and resigns in the ordinary way.

The number of A is less than

$$a = 10^{16} \sum_{N=1}^{32} \frac{64!}{(64 - N)!} = 10^{69 \cdot 7}.$$

The number of possible games is at most

$$\mu^{2\alpha+1} = 10^{10^{70 \cdot 5}} = N_{2 \cdot 185}.$$

The problem of a not too hopelessly inadequate lower bound (even a moral certainty without full proof) seems not at all easy. Unless there are a fair number of mobile men the number of positions, which dominates the top index, is inadequate; with a number of men, however, it is difficult to secure their *independent* ranging through long sequences of moves. We may consider Kings at corners protected from check by 3 minor pieces, and some Queens of each colour (9 if possible). I have thought of this question too late to try to develop a technique; perhaps some readers may compete for a record.

§11. We come now to the numbers that I describe as indirectly connected with daily life. These arise out of the enormous number $n_0 = 3 . 10^{19}$ of molecules per c.c. of gas under standard conditions, and the permutations connected with them. I will recall the admirable illustration of Jeans that each time one of us draws a breath it is highly probable that it contains some of the molecules of the dying breath of Julius Caesar.

What is the probability that the *manuscript* (as opposed to a type-script) of *Hamlet* came into existence by chance; say the probability that each of the n molecules of the ink found its chance way from an ink-pot into *some* point of an ink-line of script recognisable as the text of *Hamlet*? We can choose half the molecules in the actual ink-line to determine a narrow region into which the other $\frac{1}{2}n$ molecules have to find their chance way. If the chance for a single molecule is f the relevant chance is $f^{n/2}$. Since n is of the order of $500n_0 = 1 \cdot 5 . 10^{21}$, the crudity principle operates, and it makes no difference whether f is 10^{-1} or 10^{-10}. The adverse chance is $N_{2 \cdot 13}$.

We believe, of course, that something happened which at first sight is much more improbable; the ink came into position in orderly succession in time. But what is the additional factor? If we call the ink of a full stop a 'spot', the ink-line is made up of say $s = 10^6$ spots. We must multiply the original number by $s!$, but this leaves it quite unaffected.

(Similarly the latitude of choice implied by the italicised 'some' above makes no difference.)

§12. We all know that it is merely probable, not certain, that a kettle on a gas-ring will boil. Let us estimate the chance, by common consent small, that a celluloid mouse should survive for a week in Hell (or alternatively that a real mouse should freeze to death). Piety dictates that we should treat the problem as classical, and suppose that the molecules and densities are terrestrial. We must not belittle the Institution, and will suppose a temperature (absolute) of $T_H = 2 \cdot 8 \cdot 10^{12}$ (the $2 \cdot 8$ is put in to simplify my arithmetic).

*Let c be the velocity appropriate to temperature T, $T_0 = 280$ (English room temperature), $c_0 = c(T_0)$, $c_H = c(T_H)$. Let $\mu = kn_0$ be the number of molecules in the mouse ($k = 10^3$, say). The chance p that a given molecule has $c \leq c_0$ is, in the usual notation,

$$4\pi \int_0^{c_0} \left(\frac{hm}{\pi}\right)^{3/2} e^{-hmc^2} c^2 \, dc.$$

This is of order (the constant is irrelevant by the crudity principle)

$$p = (c_0/c_H)^3 = (T_0/T_H)^{3/2}.$$

The chance that most of the mouse has $c \leq c_0$ is not much better than p^μ. Let τ be the 'time of relaxation' at temperature T; this is comparable with the time of describing a free path; then

$$\tau_H = \tau_0 c_0/c_H = \tau_0 (T_0/T_H)^{1/2},$$

and τ is of order

$$\tau_0 = n_0^{-1/3}/c_0.$$

In a week there are $\nu = w/\tau_H$ time intervals of length τ_H where w is the number of seconds in a week.

Now an abnormal state subsides in time of order τ_H and a fresh 'miracle' is needed for survival over the next interval. The total adverse chance against survival for a week is therefore of type

$$C = (p^{-\mu})^\nu = \left(\sqrt{\frac{T_H}{T_0}}\right)^{\frac{3}{2} c_0 kwn_0^{4/3} \sqrt{T_H/T_0}} . *$$

With numerical values

$$n_0 = 3 \cdot 10^{19}, k = 10^3, c_0 = 4 \cdot 10^5, w = 5 \cdot 10^5, \sqrt{T_H/T_0} = 10^5,$$

we have

$$C = 10^{10^{46 \cdot 1}} = N_{2 \cdot 17}.$$

Of the $46 \cdot 1$, 5 comes from T_H/T_0, $5 \cdot 7$ from w, $5 \cdot 6$ from c_0, and most from n_0.

Factorizations

§13. The days are past when it was a surprise that a number could be proved prime, or again composite, by processes other than testing for factors up to the square root; most readers will have heard of Lehmer's electric sieve, and some at least will know of his developments of the 'converse of Fermat's theorem' by which the tests have been much advanced.[1] For comparison with our other numbers I will merely recall: $2^{127} - 1 \sim 10^{38}$ is the greatest known prime; $2^{257} - 1 \sim 10^{76}$ is composite though no factor is known, and it holds the record in that field; $2^{2^6} + 1 \sim 10^{19}$ has factors

$$274177 \quad \text{and} \quad 67280421310721.$$

Most numbers that have been studied have naturally been of special forms like $a^n \pm b^n$; these are amenable to special tests of 'converse Fermat' type. I asked Professor Lehmer what size of number N, taken at random, could be factorized, or again have its prime or composite nature determined, within, say, a year, and (a) for certain, (b) with reasonable certainty, (c) with luck. Much depends on the nature of $N - 1$. If a reasonably large factor or product of factors of this is known the 'converse Fermat' processes will decide the nature of N, and this for N with 50 to 100 digits. Similar results can be obtained if we can find factors of $N + 1$. Generally both $N \pm 1$ will have many small factors. The disaster of finding all three of N, $N - 1$, $N + 1$ resisting factorization must be exceedingly rare (and it suggests theoretical investigation). If, however, it occurs the value of $a^N - 1 \pmod{N}$, with, say, $a = 2$, can be calculated. If this is different from 1 then N is, of course, composite; if N *is* composite, the test, if we may judge by smaller N (of order 10^{10}) is very likely though not certain to succeed. If the value is 1, N is (accordingly) very likely but not certain to be prime. In this final case there remains, for definite proof, none but 'direct' methods, and these are applicable only up to about 10^{20}.

Professor Lehmer further tells me that numbers up to $2 \cdot 7.10^9$ can be completely factorized in 40 minutes; up to 10^{15} in a day; up to 10^{20} in a week; finally up to 10^{100}, with some luck, in a year.

---

[1] See *Math. Ann.* **109** (1934), 661–667; *Bull. Amer. Math. Soc.* 1928, 54–56; *Amer. Math. Monthly* **43** (1936), 347–354.

(Electronic calculating machines have now entered the field. The record prime is now $180p^2 + 1$, where p is the previous record prime $2^{127} - 1$. After 7 abortive tests on other numbers of the form $kp^2 + 1$ this successful one was recently made by J. C. P. Miller and D. J. Wheeler with the EDSAC at Cambridge. A converse Fermat test was used and the time occupied was 27 minutes.

The stop-press news (April 1956) is that Lehmer holds the record with the prime $2^{2281} - 1$.)

[In early 1986 the situation is rather different. With purpose-built computers, numbers of 80 digits can be factorized and numbers of 300 digits can be tested for primality. The two largest known primes are $2^{132049} - 1$ and $2^{216091} - 1$.]

*$\pi(x) - \mathrm{li}(x)$ and the Skewes number

§14. The difference $d(x) = \pi(x) - \mathrm{li}\,x$, where $\pi(x)$ is the number of primes less than or equal to x, and $\mathrm{li}\,x$ is the (principal value) logarithmic integral $\int_0^x \frac{dx}{\log x}$, is negative for all x up to 10^7 and for all x for which $\pi(x)$ has been calculated. I proved in 1914 that there must exist an X such that $d(x)$ is positive for some $x \leq X$. It appeared later that this proof is a pure existence theorem and does not lead to any explicit numerical value of X; such a numerical value, free of hypotheses, was found by Dr Skewes in 1937; his work has not yet been published, though it should be before very long. In the meantime I will report here on the matter, for there are unexpected features apart from the size of the final X.

If we denote by θ the upper bound of the real parts of the complex zeros of the Riemann ς-function $\varsigma(s)$, the famous 'Riemann hypothesis' (R.H. for short) is that $\theta = \frac{1}{2}$; if this is false, then $\frac{1}{2} < \theta \leq 1$. It has long been known that in the latter case $d(x) > x^{\theta - \epsilon}$ for arbitrarily small positive ϵ and some arbitrarily large x, so that an X certainly exists. This being so we may, for the purpose of a mere existence theorem, *assume* R.H., and my original proof did this. For a numerical X it is natural to begin by still assuming R.H. Doing this, Dr Skewes found[1] a new line of approach leading to

$$(1) \qquad\qquad X = 10^{10^{10^{34}}}.$$

In this investigation it is possible to reduce the problem to a corresponding result about a function $\psi(x)$ associated with $\pi(x)$ ($\psi(x)$ is $\sum_{n \leq x} \Lambda(n)$,

---

[1]See *J.L.M.S.*, **8** (1933), 277–283

where $\Lambda(n)$ is $\log p$ if n is a prime p or a power of p, and otherwise 0).
The 'corresponding result' is

$$'\delta(x) = \psi(x) - x - \frac{1}{2}x^{1/2} > 0 \quad \text{for some } x \leq X';$$

if $\delta(x) > 0$ for some x then (roughly — I simplify details) $d(x) > 0$ for
the same x. Next we have (roughly) an 'explicit formula' (R.H. or not)

$$(2) \qquad \frac{\delta(x)}{x^{1/2}} = -\frac{1}{2} - \sum \frac{x^{\beta-1/2} \sin \gamma\eta}{\gamma},$$

where $\beta + i\gamma$ is a typical complex zero of $\varsigma(s)$ with positive γ, and we
write $\eta = \log x$. Next, R.H. or not, a negligible error is committed by
stopping the series in (2) when γ reaches x^3.

After these preliminaries we can consider the full problem of an X
free of hypotheses, and the stages through which it went (this will involve
a little repetition).

(i) Assume R.H. Then $\beta = \frac{1}{2}$ always, and it is a question of finding
an $X = X_0$ such that

$$\sum_{\gamma \leq X^3} \frac{\sin \gamma\eta}{\gamma} > \frac{1}{2} \quad \text{for some } \eta \leq \log X.$$

The solution of this, which is a highly technical affair is given in Dr
Skewes's *J.L.M.S.* paper. There is, as explained above, a final 'switch'
from ψ to π (on well-established principles) and the X_0 arrived at is (1).

(ii) It is known that $\sum_{\gamma \leq T} 1/\gamma < A \log^2 T$ (it is actually of this
order). If, instead of R.H., we assume slightly less, viz. that no zero with
$\gamma < X_0^3$ has, say, $\beta \geq \frac{1}{2} + X_0^{-3} \log^{-3} X_0$, then $\sum_{\gamma \leq X_0^3} x^{\beta-1/2} \left(\frac{\sin \gamma\eta}{\gamma}\right)$
differs trivially over the range $x \leq X_0$ from $\sum_{\gamma \leq X_0^3} \frac{\sin \gamma\eta}{\gamma}$, and we still
have (after trivial readjustments) the conclusion '$\delta(x) > 0$ for some
$x \leq X_0$'.

(iii) It remains to prove the existence of a (new) X under the *negation* of the hypothesis in (ii). Now this negation is equivalent to the
assertion of the existence of a $\beta_0 + i\gamma_0$ distant at least $b = X_0^{-3} \log^{-3} X_0$
to the right of $\text{Re } s = \frac{1}{2}$, and with ordinate γ_0 not above X_0^3; i.e. we have
a more or less *given* $\beta_0 + i\gamma_0$ with $\beta_0 - \frac{1}{2} \geq b$. Incidentally $\theta \geq \frac{1}{2} + b$
and there certainly *exists* an X. There is now a new surprise. With

a $\beta_0 + i\gamma_0$ given as above we might reasonably expect the original θ-argument to provide an associated x with $\delta(x) > x^{\theta-\epsilon}$; alternatively, it is plausible that for *some* x the series $\sum_{\gamma \leq x^3} x^{\beta-1/2} \left(\frac{\sin \gamma \eta}{\gamma} \right)$ should exceed say 10^{-1} times the value of its individual term $x^{\beta_0-1/2} \left(\frac{\sin \gamma_0 \eta}{\gamma_0} \right)$ (and with x making the sine positive). Dr Skewes, however, convinced me that the argument does not do this: it does not deal in individual terms, and the difficulty is that any term we select may be interfered with by other terms of its own order or greater. The difficulty is not at all trivial, and some further idea is called for. In the end I was able, in general outline, to supply this.

(iv) But the problem is still not done for. Dr Skewes convinced me (this time against resistance) that in the absence of R.H. it is no longer possible to make the switch from ψ to π. This being so, it is necessary to carry through the work with the explicit formula for $\pi(x)$ instead of $\psi(x)$, with many attendant complications. This Dr Skewes has done, and his result stands at present as $X = N_4(1\cdot46)$ (improved to $N_3(3)$).[1]*

§15. The problem just considered prompts the question: could there be a case in which, while pure existence could be proved, no numerical X could be given *because any possible value of X was too large to be mentioned*? The mathematician's answer is 'no', but we do thus return to the question, with which we began, of how large a number it is possible to mention. What we want is really a function $F(n)$ increasing as rapidly as possible; what we finally substitute for n, whether 2, or u, or $N_u(u)$ makes no difference. (We must stop *somewhere* in constructing F, but one more step, say to $F(F(n))$, would overwhelm the difference in substitutions.)

We start with a strictly increasing positive $f_0(n)$. If we write $\psi^k(n)$ for the k-times iterated function $\psi(\psi(\ldots \psi(n)))$ we can define

$$f_1(n) = f_1(n, f_0) = f^{f^{\cdot^{f(n)}}} \quad \text{(to } f_0(n) \text{ indices, say)},$$

where for clarity we have suppressed the zero suffixes in the ascending f's.

This defines an increase of suffix from 0 to 1; we suppose f_2 derived similarly from f_1 (in symbols $f_2(n) = f_1(n, f_1)$), and so on. We now take a hint from the notation of transfinite ordinals, and form

$$f_{f_{f(n)}}(n)$$

[1][The relevant paper of S. Skewes was published considerably later than anticipated: *Proc. London Math. Soc.* (3) **5** (1955), 48–70.]

(to $f_{f(n)}(n)$ suffixes, say). We can now say: scrap the existing definitions, as scaffolding, and define *this* to be $f_1(n)$, and carry on as before. We can scrap again, and so on: here I decide to stop. Once we stop we may take $f_0(n) = n^2$, or $n + 1$ (*what* we take does not matter provided only $f_0(n) > n$).

The reader will agree that the numbers mentioned are large: it is not possible to say *how* large; all that can be said about them is that they are defined as they *are* defined. If it were desired to compare terms in two rival systems a considerable technique would have to be developed.

LION AND MAN

'A lion and a man in a closed circular arena have equal maximum speeds. What tactics should the lion employ to be sure of his meal?'[1] Ideally, the reader should spend 24 hours' honest thought before proceeding further. Experience shows that the enjoyment of what follows is spoilt for the casual reader.

It was said that the 'weighing-pennies' problem wasted 10,000 scientist-hours of war-work, and that there was a proposal to drop it over Germany. This one, invented by R. Rado in the late thirties swept the country 25 years later; but most of us were teased no more than enough to appreciate a happy idea before arriving at the answer, 'L keeps on the radius OM'.

If L is off OM the asymmetry helps M. So L keeps on OM, M acts to conform, and irregularity on his part helps L. Let us then simplify to make M run in a circle C of radius a with angular velocity ω. Then L (keeping to the radius) runs in a circle C_L of radius $a/2$, touching C, at P, say, and M is caught in time less than π/ω. This follows easily from the equations of motion of L, namely $\varphi \to \dot{\theta} = \omega$, $\dot{r}^2 + r^2\omega^2 = a^2\omega^2$. A more geometric way of proving this is indicated in Figure 21. The angle φ subtended by the arc MM' at O is precisely the angle subtended by the arc LL' at O. Since O is the centre of the circle C of radius a and is on the circle C_L of radius $a/2$, the arcs LL' and MM' have the same length. Thus if L keeps to the radius OM then L runs along C_L.

It is, however, instructive to analyse the motion near P. For this

$$\dot{r} \geq (a - r)^{1/2}/K, \quad \text{and} \quad t < \text{const.} + K \int (a - r)^{-1/2} dr.$$

The integral converges (as $r \to a$) with plenty to spare — plenty, one

---

[1]The curve of pursuit (L running always straight at M) takes an infinite time, so the wording has its point.

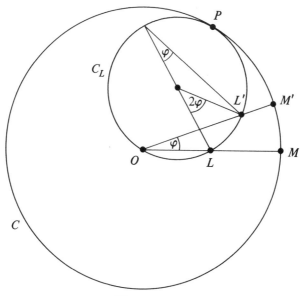

Figure 21

would guess, to cover the use of the simplifying hypothesis. At this point, the reader should stop again in order to digest the argument above.

*The professional will easily verify that when M spirals outwards to a circle, and, with obvious notation (ω varying), we write $x = r_M - r_L$, we have $-\dot{x} \geq (\omega^2 r_L)(x/\dot{r}_L) = X$, where $X/x \to \infty$. Then $t < \text{const.} + \int_x X^{-1} dx$, and in this the integral increases more slowly than $\int_x x^{-1} dx$: it is a generally safe guess in such a case that the integral *converges*.*

All this notwithstanding, the 'answer' is wrong, and M can escape capture, (no matter what L does).[1] This was discovered by Professor A. S. Besicovitch in 1952; here is a simple version of it.

I begin with the case in which L does keep on OM; very easy to follow, this has all the essentials in it (and anyhow shows that the 'answer' is wrong). Starting from M's position M_0 at $t = 0$ there is a polygonal path $M_0 M_2 M_2 \ldots$ with the properties: (i) $M_n M_{n+1}$ is perpendicular to OM_n, (ii) the total length is infinite, (iii) the path stays inside a circle round O inside the arena (Fig. 22). In fact, if $l_n = M_{n-1} M_n$, we have $OM_n^2 = OM_0^2 + \sum^n l_m^2$, and all is secured if we take $l_n = cn^{-3/4}$, with a suitable c. Let M run along this path (L keeping, as agreed, on OM). Since $M_0 M_1$ is perpendicular to $L_0 M_0$, L *does not catch M while M is on $M_0 M_1$*. Since L_1 is on OM_1, $M_1 M_2$ is perpendicular to $L_1 M_1$ and L does not catch M while M is on $M_1 M_2$. This continues for each

---

[1]I used the comma in line 10 p. 114 to mislead; it does not actually cheat.

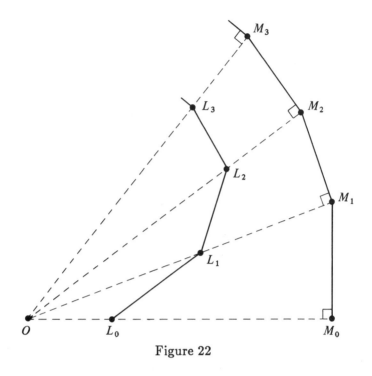

Figure 22

successive $M_n M_{n+1}$ and for an infinite time since the total length is infinite.

I add a sketch, which the professional can easily complete, of the astonishingly concise proof for the quite general case. Given M_0 and L_0, M 'constructs' the polygon $M_0 M_1 M_2 \ldots$ described above, but runs along another one, $M_0 M_1' M_2' \ldots$, associated with it, but depending on what L does (Fig. 23). $M_0 M_1'$ is drawn perpendicular to $L_0 M_0$, N_0 is the foot of the perpendicular from O on $M_0 M_1'$, and M_1' is taken beyond N_0 from M_0 so that $N_0 M_1' = l_1$. If L is at L_1 when M is at M_1', then $M_1' M_2'$ is drawn perpendicular to $L_1 M_1'$, and M_2' is taken on it so that $N_1 M_2' = l_2$; and so on. Clearly $OM_n'^2 - OM_{n-1}'^2 \le l_n^2$, $OM_n'^2 \le OM_n^2$, and the new polygon is inside the same circle as the old one. Since $M_{n-1}' M_n' \ge l_n$, the new polygon has again infinite length. And as before L fails to catch M.

[The problem has several variants. H. T. Croft (*J. London Math. Soc.* **39** (1964) 385–390) considered the case when M has to run along a path of uniformly bounded curvature (L can catch M, although if L stays on the radius, M can escape). Croft showed also that n lions can catch a man in an n-dimensional ball although $n-1$ cannot. One of the

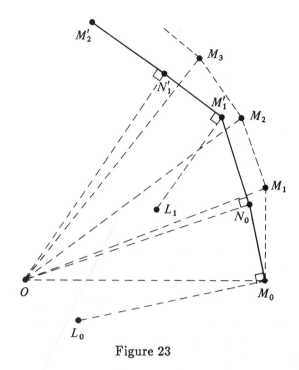

Figure 23

many unsolved 'lion and man' problems is the following. Can two lions catch a man in a bounded area with rectifiable lakes?]

PEOPLE

Random Jottings on G. H. Hardy

About 1910, Hardy, Norton, and I were talking together and got on to the subject of writing and style. Hardy said: 'What's wrong with my style is not lack of expressiveness or lucidity, but vulgarity.'

He said to me in later life that he didn't like his youthful style, but his calm estimate in 1910 startled me. He had in those days a great reputation for general intellectuality (and among half-enemies for arrogance). He read a notorious paper to the 'Sunday Essay Society' (a high-brow Trinity Dons' society) in this vein. I found it after his death with the additional comment: 'Nothing of any real interest.'

He had some odd limitations (in unimportant ways). He *said* he never knew the actual Curate's Egg story, though he recognized the context in which references to it were made, and he said he once angled to find out the actual thing.

His spelling was not immaculate, odd for anyone of his time (we could all spell all but perfectly from the age of 12). And he once queried whether when someone who wrote 'depreciate' 'didn't really mean "deprecate" — there is a word deprecate isn't there?'

He was not good at crosswords, though he did them on holiday.

Our habits were about as opposite as could be. He was unhappy except with bright conversation available, and his best time for work was the Long Vacation, in Oxford or Cambridge, with tennis or cricket in the early afternoon. He preferred the Oxford atmosphere and said they took him seriously, unlike Cambridge.

He took a sensual pleasure in 'calligraphy', and it would have been a deprivation if he didn't make the final copy of a joint paper. (My standard role in a joint paper was to make the logical skeleton, in shorthand — no distinction between r and r^2, 2π and 1, etc., etc. But when I said 'Lemma 17' it stayed Lemma 17.)

A ms. page of 'A maximal theorem with function theoretic applications' (1930); the main text is in Hardy's beautiful handwriting, the additions are Littlewood's.

He would copy an argument, in which he was interested, in longhand. He thought no more of buying a stack (of the highest quality) of 'scribbling paper' than I would a bottle of sherry. For a long time he wrote in ink, though he did later switch to pencil except for the final mss. If the first half-sentence went wrong, he rejected the sheet and took another.

He always said himself that he liked to pose as Oxford at Cambridge and vice versa (he said this in America, where he was doing the like for Harvard vs. Princeton or vice versa). At *cricket* he kept a passionate

Cambridge loyalty. He told me once he had mellowed; he still wanted a Cambridge victory by innings and 200, but no longer wanted the Oxford Captain to be hit in the stomach by the fast bowler. His hatred of rowing made him back Oxford when he was at Cambridge and vice versa.

He was indifferent to noise; very rare in creative workers, at least when no longer young. (At 20 I could work in a room with a crying baby; now I fret if someone coughs in the room above.) Rosetti said of dogs, '*Why* do they call them dumb animals?' There is an atrocious dog around my Davos hotel (I once told the concierge I would pay 100 francs if it happened to disappear).

Until he was about 30 he looked incredibly young: there was a legend that when he was a Prize Fellow the Kitchen Office wouldn't let him order beer; it was not allowed to Scholarship candidates.

In my first year, Dons lunched at the same table as undergraduates, and I once innocently happened to sit next to a block of them. Presently I heard what was apparently an undergraduate chaffing the infinitely venerable Henry Jackson, with great elegance and verve on both sides.

I never knew about his not having a looking-glass anywhere, until it came out to see his own face. Everyone thought he looked appropriately striking. Bertrand Russell said to me (*c.*1911) that he had the bright eyes that only very clever men had ('clever' then meant more than now). *I* would have said his face (until perhaps his premature old age) was beautiful.

Another contrast with me. Hardy liked to end the day's work feeling that the last idea had some hope in it. To me it is most repugnant to be enjoying possibly false hopes; I want if possible to know the worst. (As a matter of fact, those crises happen at the end of the day, and especially the one before a day off, with diabolical frequency.)

He must be chuckling in Heaven over my possessing the finest mathematical library in the country.

He had violent prejudices against food; his vocal obsession with the horror of hot roast mutton was something of a joke.

He believed that the taste for Old Brandy was a pose. An 'old brandy stunt' began by meaning a dishonest pretension to accepted highbrow taste. By extension it came to cover aggressive opposition to it; Samuel Butler's 'Handel is better than Beethoven' is a typical o.b.s.

S. Skewes. Invited by his tutor (Arthur Berry of King's) to lunch, he considered his letter of acceptance. He rejected 'shall come with pleasure' etc., because he *had* to go. Similar difficulties of sincerity barred all other forms he could think of, so he made *no* answer. When he arrived and was told 'it is usual to answer an invitation', he could again think of no sincere answer, and was too inarticulate to explain. Twenty years later he told me all this; he would then have been in command of the situation, but in the meantime Berry had died.

Sir James Fraser. Before 1914 he went annually to Scotland, accompanied (as most of us used to be then) by a large book-box. One year he was told that his luggage was overweight, and that he had to pay 5 shillings or so. He paid, but wrote to the Company questioning the matter, and saying that he had done the same thing for over 20 years and had never been charged. They quoted regulations and convinced him. He then calculated what he owed them over 20 years and sent a cheque. They replied that they could not think of accepting it. He then wrote: the money is not yours to refuse, it belongs to the shareholders; you *must* take it.

P. V. Bevan. Bertrand Russell once asked me what the French word for some tree or other was in English. I said I didn't see the use of knowing the French for a thing I wouldn't know in English. He replied 'Not at all; Bevan knows the word for woman in at least 14 languages, but he wouldn't know one if he saw one.' (The response was immediate, but perhaps he had said it before.)

A. L. Smith, the Oxford historian and later Master of Balliol, visited Trinity in 1913, and came to tea with Russell, Bevan and H. T. J. Norton, and others present (Russell told me about it the next day or so). He was pontificating *á la* Oxford 1913 and A. L. Smith, but ran into a succession of hideous rebuffs. I remember only two (but there were more). The one: 'How sad it was that the Palace at Peking had been burnt the year before' — to which Norton responded: 'Well, I don't know; it was there 6 weeks ago when I was visiting it.' The other: he quoted, in 'Persian', a proverb. Bevan exploded — to the effect that what he said was nonsense, and should be something else, which further was not Persian.

F. J. H. Jenkinson, the University Librarian before 1914. He said one day at lunch 'I've had my entire morning wasted by *The Times.*' 'How's that?' 'That 70 page South American supplement.' 'Oh! I didn't read it.' 'Didn't *read* it; you must be terribly strong-minded.'

Neder. His mathematical work consisted of a number of small observations, just worth noting by his betters, beautifully written out in case anyone should want to bother with the details (which they didn't). As a man he was a superb physical specimen, and left a trail of illegitimate children wherever his various posts took him.

Chike Obi. He took a London B.A. from West Africa, mathematics not a subject. He wrote asking for the mathematical books he wanted to learn, including Poincaré on Celestial Mechanics, and taught himself. He managed to get accepted in Cambridge for a Ph.D. (turned down in London), and then struck a modest, but quite genuine, vein of oil (making some experts on differential equations look quite silly). He had no idea whatsoever how to write, and I had him for a year — one term I had him for 90 minutes a day every other day. (Each session began by my removing trumpet parts from the orchestration.) He had a curious belief that he could get something out of nothing by a transformation $x = \epsilon y$ and *then* using the fact that epsilon was small. I finally said this 'was a belief in ju-ju' (which went off all right, if rather daring). I also said he alternated classical off-drives with pure cow-shots, and got an understanding grin.

There was a film about this time about an intellectual negro going back to Africa and all but succumbing to a witch-doctor. The English Resident whispered in his ear when he lay in a coma 'Africa needs you'. Obi was the image of the — witch doctor. ('Obi' *means* witch-doctor'.)

Lindemann (Lord Cherwell). In the '20s I had some complex about him (for no reason I can conceive) and could never remember his name. I adopted the mnemonic that it was the same as that of the man who proved π transcendental. The result was that I could not remember *his* name either.

Lindemann's Chess problem. Obviously insoluble because of stalemate. The key-move, however, is Q.P. to Q.8 becoming *black* knight. (White moves first.)

Oscar Browning = O. B. (I lunched with him in 1905 and he introduced me to late Beethoven, on a pianola.) 'After all, a microbe is only a little O. B.'

'Put out my second best dress suit; I am dining with the King of Greece.'

Caught naked after bathing by a party of ladies, and with only a handkerchief available, he covered his face. 'Anyone in Cambridge would recognize my face, no-one would recognize my —.'

'The other Browning.'

Montagu Butler. (The last appointment of a Master of Trinity by the Crown not wished by the College. Doubtless advice was taken, but their idea was some Senior, politically active, who was quite out of touch with real opinion.) A. N. Whitehead always referred to him as 'that fathead'. Henry Jackson, as 'that —'.

Adam Sedgwick had a theory that his apparent feebleness and fat-headedness was all pose, and had a story. He was waiting downstairs in the Lodge for an interview, when he heard a manly shouting and virile descent of the stairs. This was checked by the butler's announcing a visitor, and presently the door opened slowly and Montagu Butler crept in.

E. Harrison. Once in Hall (*c.*1927) he asked me the meaning of 'bleak'. On seeing G. E. Moore opposite going into a giggle, I thought: after all, why not?, and said: 'Well, Harrison, if you really want to know, *you* are the typical bleak man.' 'Not dim, I hope?' 'No, not dim.'

This story has now become well-known, and Hardy and Macdonald had a discussion on it, concluding that my remark was just the straight, intelligent, and civilized thing, with which I now agree.

Ever after Harrison treated me as a pet. Once he said: 'I do wish you would answer the Vice-Chancellor, he has written to you four times.' 'They must have been in those envelopes for the notices that one's salary has been paid into the bank, and went into the waste paper basket.' 'All right, I'll tell them to put them in scented envelopes marked "Private".'

Earlier, when I was a College Lecturer, he agreed that I could leave for Cornwall on May 29. There was no *real* reason why not, other than red tape, and he waived it.

He once made me walk with him on the grass of the Avenue, to renew his childhood's pleasure in shuffling through dead leaves.

At the College Meeting about the very controversial lighting of the Hall: Pendente lite. (A case of discipline, Mr Tritton having persistently blown a hunting horn at 2 a.m., was being considered by the Council. Harrison passed round a note: And hear old Tritton blow his wreathed horn.)

During the First World War he was a naval officer and shaved his moustache. On visiting Cambridge the Master (Montagu Butler) asked him at a dinner whether he was related to 'our dear Ernest Harrison'. Adopting a certain philosophical view of 'relations' (repudiated by Russell) he replied: No.

He had a brother in comparison with whom he was gentleness itself. (His first words to a tutorial pupil: 'Be brief.' But he took infinite trouble with them.)

A. E. Housman. I remember telling him that someone had just told me of an enthusiast for the Boy Scout Movement, 'The most unlikely man you could think of' (actually H. F. Baker). I said that he was not *my* choice of the most unlikely. He gave a reluctant grin, but a grin.

The Ellises gave him a dinner (rook pie). Later I heard Polly protesting to her husband. 'I *didn't* kiss him, I only stroked his face.' I repeated this to Housman in our annual correspondence — he used to borrow my rooms for his 'family dinner party' to sit in after dinner; an admitted outrage, but he gave me first editions of his books.

I similarly once repeated a remark made by F. H. Sandbach (extremely odd, but clearly genuine, reaching me via Polly): what the College wants is more people like X and Y (= A. E. Housman and J. E. Littlewood).

G. H. Hardy sat in Hall between Housman and me, and oblivious of his presence began talking about Housman's Inaugural Lecture: 'A bit on the cheap side, of course;' at which moment, to my eternal shame, I nudged Hardy and brought him to a stop. If he had been allowed to go on he would obviously have said something flattering.

A. E. Housman had our Book Club anniversary dinner shifted from 'The 1st Wednesday in March' because that was Ash Wednesday and he wouldn't miss the salt cod. He designed the dinner, which was a procession of Hardy's *bêtes noires*. Sherry instead of the cocktail he wanted. Oysters, which he never touched, with double Stout (i.e. 'double' the 'hot roast mutton' in drink). Turbot, the h.r.m. in fish, followed by roast saddle of mutton.

I once said to him in Hall: 'Suppose there was a poet, Shakespeare combined with Milton, and 6 inches high; wouldn't you patronize him?' He said that the temptation would be too much for him.

A. H. F. Boughey. He was a double first in Classics and Mathematics. People visiting Trinity at a certain period used to ask: what has happened to that brilliant man Boughey, everyone now seems to be talking about a man they call J. J.?[1]

Before 1914 the High Table dinner had choice of soups, fish, entrée, choice of joint, e.g. sirloin of beef and pheasant, sweet, choice of cheese or

---

[1] J. J. Thomson, the great physicist; later Master.

savoury. (Also two sorts of sherry in decanters, claret *and* burgundy in decanters, all ad lib — unlimited 'High Table whisky' and Perrier water if you preferred it.) Boughey had all the courses. At the joint course (and the helpings were very generous) he had, first a 'little more sirloin on the same plate', and *then* the pheasant.

Being lame he went about on a tricycle, and perhaps he did take some exercise on it.

E. Landau. He possessed an enormous capacity for work — up to 12 or more hours a day. He was also an enormous eater, but took 80 minutes siesta after lunch. Normally Landau took no exercise, but could on occasion walk 20 miles. He was extremely timid physically (e.g. in the gymnasium, where it was once customary for *everyone* to perform *something*). He worked to completely rigorous rules. We were once 'working' together in Cambridge, and started immediately after breakfast. I presently said: excuse me for a minute or two. 'Two minutes 47 seconds.'

He had great integrity, and fundamental modesty. It was said round 1912 that it gave him the same pleasure when someone else proved a good theorem as if he had done it himself. He carried the duty of stamping out bad mathematics perhaps to excess. Just before the First World War he had written a scarifying review of a piece of work by a Frenchman. The proofs arrived during the war, and he said to himself: 'This passage will be attributed to political prejudice,' and deleted it. At the final proof he said to himself: 'What I wrote I wrote with due consideration; whatever the appearances, if I suppress the passage, I *shall* be acting with political prejudice;' and back the passage went.

We were walking in London and arrived at the Albert Memorial. On the way I threw out something of the high-brow reaction to the work. When we all arrived all he said was: 'I *like* that.'

He was completely non-musical (as were Klein and Hardy).

He had a number of *mots* to his memory. Asked for a testimony to the effect that Emmy Noether was a great woman mathematician, he said: 'I can testify that she is a great mathematician, but that she is a woman, I cannot swear.' (She was very plain).

When G. H. Hardy wrote after the First World War to the effect that he had not been a fanatical anti-German, and felt confident that Landau would wish to resume former relations, Landau replied: 'As a matter of fact my opinions were much the same as yours, with trivial changes of sign.'

There used to be a 'Carathéodory constant' for a problem in the Bloch-constant complex. This was later proved to have the value $\frac{1}{16}$, and Landau commented, in a review of the subject: 'I cannot call this constant "Carathéodory" because it already has another name, namely $\frac{1}{16}$.'

There are a couple of grim stories about his treatment of his Privat Dozents. (In his defence, he simply did not know what it was like to be tired.) One was that when the man was recuperating in a hospital, Landau climbed a ladder and pushed a chunk of work through the window. The other is about the enormously exacting job of correcting the proof-sheets of his three-volume lectures on number theory. (He claimed for many years that there was no misprint in his work, though he did in the end lose his virtue. He read proof-sheets 7 times, once for each of a particular kind of error.) The Privat Dozent had at this moment just got married and was to start on his honeymoon. Landau's first idea was that the proofs were to be taken along. There was general distress, and ultimately Landau's wife, the only person who could influence him in the last resort, was thought to have succeeded. However, Landau went up to the Privat Dozent and said: 'It's all right, you need only do one volume.'

J. G. van der Corput. The Lehmers invited me to lunch in Berkeley, and van der Pol and van der Corput were there. All four of them are extremely charming, so we were at once on the easiest possible terms.

Van der Corput is, I should say, unique in never taking a holiday. He finds his work pleasant and effortless, does not suffer the usual agonized frustration when stuck, and is just as pleased when someone else does the thing instead of himself. All this is the more remarkable when one thinks of the extreme complexity and weight of his work.

W. H. Young. On sea voyages one has one's initials on one's deck-chair. His had: W. H. Y., Sc.D., F.R.S. Told me by Scholfield, who went to India with him. It is a fair light on his fantastic character that even if this story is not true, it is perfectly credible. (He was, of course, far above a mere F.R.S.)

G. D. Birkhoff (elder). When he met you he opened: 'Sir, you are the greatest mathematician in the world.' To which you reply 'With one exception.'

A woman[1] once told me that I was reputed to be the rudest man

---

[1] A *man* wouldn't.

in Cambridge. I wonder. My own idea is that I am always the soul of courtesy.

Moore and Russell (c.1876?) were having a philosophical discussion in Hall. Russell suddenly said: 'You don't like me, Moore, do you?' Moore replied 'No.' This point disposed of, the discussion proceeded as before.

R. V. Laurence. The day he ceased to be Junior Bursar (and Pooh-Bah) he got a leg of chicken in Hall.

Miss Cartwright. She is fundamentally very modest (though dutifully asserting her rights as a representative Academic woman). In earlier days she used to make, in blandest tones, the most devastating remarks, almost all of which landed on Hardy. When he returned from Oxford she said something which on the face of it suggested that Hardy was a good second-rater. And she kept this up. When there was a question of a mathematical dinner and I suggested that Miss Cartwright should sit next to Hardy, he said: 'Well; but her fast ball is so very devastating.'

I once said to Trevelyan (who asked about her): 'One said: "Perhaps I may be stupid, but ..." ' She would reply: 'Well, yes, I think so.'

A. S. Besicovitch and the Russian temperament. He will not play my Patience *and* keep records, because it would mean he could do nothing else.

On completing the Moderate climb at Pedn y Vounder he sat triumphantly down, without looking, naturally, on a gorse bush.

When a Russian feels he must go out and dance naked in Great Court, he just has to. Besicovitch confided to me that when he first came to England he thought we were poor creatures in this way; but he came to think we were happily integrated.

A. Ostrowsky. After a period of surprisingly easy relations a chill came, for no good reason (though I had a theory). I asked Besicovich whether his experience was the same. 'Oh, yes.' 'Do you know why?' 'Oh, yes.' 'How?' 'Well, Ostrowsky and I were taking two ladies to the cinema. Ostrowsky's lady asked him: "Professor Ostrowsky, what is your subject exactly?" "You should ask Professor Besicovitch that." I said: "The fact is I have never read any of Ostrowsky's work." '

A. W. Verrall. It was the custom (*c.*1905) to read the roll at lectures (in alphabetical order). Verrall came to Mr Shufflebottom, Mr Sitwell, burst into his crow of laughter, and never read the roll again.

At a Scholarship Examination, Dykes pointed out to me that the list had the consecutives Alchin and Alcock.

Miscellaneous recollections about Bertrand Russell.

He told me that writing gave him no trouble at all. As soon as he knew what he wanted to say, the words came as easily as in an unexacting conversation.

He had a secret craving to have proved *some* straight mathematical theorem. As a matter of fact there *is* one: '$2^{2^a} > \aleph_0$ if a is infinite'. Perfectly good mathematics.

(This weakness is very common with people who take the Mathematical Tripos and then switch, e.g. John Maynard Keynes, to some extent F. P. Ramsey, and possibly R. B. Braithwaite.)

He said that what Kant did, trying to answer Hume (to whom I say there is no answer), was to invent more and more sophisticated stuff, till he could no longer see through it and could believe it to be an answer.

That every argument of Hegel came down to a pun, (often involving the word 'is').

He told me (*c.*1911) that he had conceived a theory that 'knowledge' was 'belief' in something which was 'true'. But he met a man who believed that the Prime Minister's name began with a B. So it did, but it was Bannerman and not Balfour as the man had supposed.

He said that *Principia Mathematica* (for which he did all the dirty work, since Whitehead was a hard-working lecturer) took so much out of him that he had never been quite the same again.

Lulworth 1919. We spent the Long Vacation together at Newlands farm, with guests from time to time. Some noteworthy things happened. I had just been reading A. S. Eddington's Report on Relativity (Eddington had not before discovered his flair for brilliant journalism and the impact was all the greater). I felt the theory was about the greatest intellectual advance and illumination that had ever happened. I explained it to Russell, who at that time knew no physics. He was similarly staggered. Suddenly he burst out (to Dora's consternation): 'To think I have spent my life on absolute muck.'

We were of course waiting for the results of the Eclipse test of Relativity. I wrote to Eddington imploring information as soon as possible. Russell remembers a telegram; I remember only a very non-committal letter.

I had to go away for a painful two days at P. E. B. Jourdain's death-bed. Jourdain had for some time thought he had a proof of the Multiplicative Axiom (or Axiom of Choice), and would have died happy if it were accepted. When discussing with Russell the difficulty of dealing with this, I rashly opened my mouth with a suggestion, and in the end it was I (and I think Dorothy Wrinch) who paid the visit.

Jourdain expounded his latest version verbally, and I took the line that there was a new point involved which I should have to think over carefully, for say a day. Jourdain instantly said: 'My dear man, you know perfectly well you can tell in 10 minutes whether a proof is right or wrong.'[1] I brazened it out — only thing to do — and I suppose all went as well as one could hope. It was no good just lying (and finding a mistake if after all he did recover) because of his wife. It turned out she didn't care a damn about his intellectual life.

One day we were visited by a 'single-tax' enthusiast who wanted a talk with Russell. His wife wrote poetry, and we sampled a sonnet. This was addressed to 'Him who sleeps by my side' and contained the line 'Too full of the world's meat and wine'. We accordingly provided a slap-up lunch with lots of drink; to find he was a teetotal vegetarian.

Alan Wood, writing his book on Russell, came to have a talk with me (at Russell's instigation) to verify various recollections, especially about Lulworth. Russell had told him the above story. I said it was a wonderful story, but I hadn't the faintest recollection. But two days later it came back to me that all the story was exactly true. (Shows what one can forget and what one must lose by not keeping a notebook.)

Russell's final election to a Fellowship was an odd business. I went and interviewed G. M. Trevelyan, the Master, about electing him to an Honorary one. There was a lot of reluctance: moral character, what will parents think? He did grin when I said the objection was to his principles and not to his actual practice.

---

[1] He himself became rhetorical and emotional at the fallacious point.

Not long afterwards, D. A. Winstanley met me in the King's Avenue, stopped me, and said: 'You will be pleased to hear we've elected Russell an ordinary Fellow.' The (apparent) moral distinction made by the Council calls for record.

Before 1911, Russell wore a bushy moustache; he looked incredibly mild, rather dim, so much so that one had an impulse to patronize him. When he first appeared shaved it was impossible not to show shock, but he was already accustomed to this. He said, however, that there was a class of people who noticed no difference. An extreme case of people seeing things differently.

He said once, after some contact with the Chinese language, that he was horrified to find that the language of *Principia Mathematica* was an Indo-European one.

The Moose story. Russell was giving tea to a dim American mathematical Professor and his wife, G. H. Hardy and myself being the rest of the party. The wife said: 'You should tell them the story about the Moose.' Our worst forebodings were fulfilled, and the dreadful thing crawled to its miserable end. Hardy and I gazed fixedly at the floor. After a pause, Russell: 'HA' (with strong overtones of ridicule). Then: 'HA-*HA*, HA=HA=HA...' (in his staccato genuine laugh). This set Hardy and me, and finally Russell himself, into hysterical laughter. The story was a terrific success, and it is sad to think of later sufferers.

Bertrand Russell and I each met Einstein once. I was put next to him at the dinner at Trinity to Baldwin when he became Chancellor, on the dubious ground that I could speak German — Einstein then spoke no English. He was very good at speaking simply, and we managed. We talked mostly about music. Einstein 'liked' Baldwin when he spoke. I asked him if he 'liked' Lord Sumner when he spoke (he made an angry scene and tore up his notes) — Einstein hedged. Opposite me was the American Ambassador, pale with anxiety about his speech (which turned out to be very odd). On his left the Duke of Gloucester, and on his right, a sporting peer. These were all honorary graduands, created by Baldwin. The peer got considerably drunk and heckled Einstein across the table — Einstein could not understand a word. The peer boasted afterwards of having stumped Einstein on Relativity.

All Bertrand Russell said about their meeting (apart from how wonderful Einstein looked, and his wonderful eyes) was that Einstein told him a dirty story. I said that Einstein was well known for consummate tact in adapting himself to his company.

He once spent the night in my rooms when I was away. I returned to find that a half-full jar of barley-sugar (which I take with tea at breakfast time) contained *one* unit. There are conventions on this point. A man of ungovernable passions.

He says in *My Philosophical Development* that after a mild attack of 'flu in his second year at Cambridge he was no good for work for several months, and his mathematical performance suffered. This is the first time I, and I should guess anyone else, had heard of this — probably he didn't suspect the connexion at the time. I similarly did not suspect what were probably very severe after-effects of 'flu some time before I took the Scholarship Examination and got only a Minor Scholarship. But my attack was very severe and sudden (temperature 105° at 11.00 a.m., and in bed for 10 days, with a nurse). I gather he never suffered after-effects like this again. Nor, I think, did I, until 1960. My attack (in Switzerland) then was so mild that I never thought of going to bed, and took walks. But the after-effects, for work, lasted at least three months and were severe.

ACADEMIC LIFE

My Little-go.[1] For Part I, Latin and Greek, I worked quite hard, having had an inadequate education in South Africa. (When I went to St. Paul's I was put in the 'Special' for about 4 weeks (it seemed an eternity) to learn some Greek (mostly conditional sentences about crocodiles). A Fellowship-class man who had suddenly to learn Greek for the purpose (e.g. A. S. Eddington and S. Chapman) could manage in 6 weeks. The record was held by Hirst, of my first year up, who did it in 3. He memorized the English of the Greek Testament, and translated one passage he hadn't reached by counting words from the last instance of $\beta\alpha\pi\tau\iota\zeta\epsilon\iota\nu$ (fatal to go too far). The set book had to be taken seriously: I think he worked 12 hours a day. Irregular nouns and verbs he memorized easily.

Part II was Mathematics *plus* 'Paley's Evidences',[2] *plus* an English essay. I did read some of P. but, rightly or wrongly, found it boring. I got through by reading 'Paley's Ghost' intensively for 30 minutes just before the paper (rejecting anything that would be difficult to memorize). The Essay was on 'Julius Caesar' (Shakespeare) as set book (I had 'done' this in the Cape Town Matriculation). For this and the mathematical papers I made a point of coming out at 30 minutes, the minimum allowed. It would have been no trouble to take Part II if I did fail.

False Modesty. One evening of 1953 in the Combination Room H. A. Hollond (presiding) remarked to me that it was 50 years since we came up together for the Trinity Scholarship Examination. I mentioned in fact I only got a Minor Scholarship. Something he then said called for

---

[1] An examination held four times a year which one had to take before Matriculation. Its official name was 'The Previous Examination'.

[2] 'View of the Evidences of Christianity', a treatise by William Paley, written in 1794, demonstrating the truth of the Christian religion. The candidates prepared from a précis of the volume.

the admission that this was not really normal (I was suffering the after-effects of severe 'flu), and normality returned with a Senior Scholarship in March. Between us was sitting a dim Dean of a dim College. It only occurred to me later, and Hollond said he noticed nothing, but the dim Dean found my remark intolerably arrogant. A mysterious query about the candidates of today and arrogance I realized later to be an outraged protest. There is much less false modesty among dons than elsewhere — they break some conventions of ordinary life. I remember noticing this, and, after the first shock, with approval, in Hardy's conversation about 1912.

My first lecture. Trigonometry for the better third of the Intermediate Ordinary Degree class at Manchester. Horace Lamb, the Professor at Manchester and a Past Fellow of Trinity, had given me the usual, and desirable advice for a bright young man from Cambridge: 'You can't be too simple.' I had prepared a careful approach to the idea of an angle ($\angle$ is a small one, $\langle$ is a large one). It soon became evident that there was unrest and amusement. 'You know all this?' 'YES.' 'Do you know sin and cos?' 'YES.' Having grasped the idea I said 'Do you know the general proof of $\sin(a + b) =$ etc.?' 'No.' 'Very well', and off I went. If you do the thing unawares (and I hadn't dealt with it since early schooldays) you find at a certain point you ought to have begun with a preliminary 'lemma'. There was a pause. A slight shuffling of feet was growing. I had visions of a promising academic career cut short at the outset, any parent's grey hairs. At that moment the bell rang. 'We will do this properly next time.'

A first lecture is inevitably terrifying. A mathematician can't *read* a prepared script, and has to leave a lot to the inspiration of the moment. He can't know whether he won't find himself tongue-tied.

The first lecture of a new year renews for most people a slight stage-fright — I remember Hardy agreeing with this. The real old hand is not stage-frightened at anything of a familiar academic type (Lecture abroad, paper to the Royal Society, etc.). But I remember that a 'talk' in the top form in mathematics at St. Paul's was so unfamiliar that there was an initial stage-fright.

In my first 4 years as lecturer I did 9 Scholarship Elections. There was then a Senior Scholarship one in March, with 3 examiners in Mathematics, and I did all the dirty work. The odd one was for the December before I was appointed (on approval). I was the only Trinity man at the first meeting, Chairman C. Smith, Master of Sidney (S. L. Loney and J. Edwards, who wrote dreadful books, were also Sidney men), and there were other senior men. W. L. Mollison put up a proposal to abolish the 'Essay paper' which was a Trinity paper (later abolished). I felt I must go down fighting, and said I couldn't agree without consulting any colleagues. To my surprise the thing was immediately dropped (can't think why they tried it on).

A. R. Forsyth wrote a C.U.P. book on functions of two complex variables. It is a thoroughly bad book. On skimming through much that was 'Ph.D. variation' on one variable, I struck one significant theorem. Ah! After 10 minutes I found it was fallacious.

The referee said it was not acceptable, but the Press considered they could not refuse to publish a book by a Professor of the University. This was about 1912. Much later they turned down my 'Mathematician's Miscellany'.

G. T. Bennett (Senior Wrangler beaten by Phillipa Fawcett) once said to me: 'I've been working out how to "untear up" a torn up ms. There are two schools, whether you put the right hand tear on top of the left or vice versa) apparently equally divided.' He worked out appropriate formulae.[1]

Between the first and second days of a Fellowship Election I went up to London, and was rather early for the train. One of the candidates, a former pupil for Part I, came up and talked to me. I never hold inarticulateness, or even what seems dull stupidity, against a candidate (there are examples!), but there is a sort of fluent asinity about which there is no doubt; I was quite clear that he was intellectually impossible for a Fellowship, no matter what anyone said, and I voted solidly against him. He was elected, and after the subsequent dinner C. D. Ellis came up: I want to introduce X. *It was a totally different man.* We went on to identify the other man. Ellis said 'Of course you're perfectly right about *him.* Unless you have had the experience it is impossible to realize what a shock it was. The man was at his last shot, and if he failed I should probably never have known my mistake. If I *had,* I should have confessed to the Council and recommended another chance. The trouble was that I 'knew' X was the man, and checking never entered my mind.

Meeting a research pupil after the Michaelmas Vacation, and finding he had taken a complete holiday, I said, more or less, 'Holidays are very important — if you feel like 6 weeks off, well and good, and anyhow the

---

[1]From 1882, women also took the Mathematical Tripos; they were not listed in the strict order of merit, but their positions were given as 'equal to 31' or 'between 25 and 26', say. Before 1890, when P. G. Fawcett, a cousin of Littlewood, was ranked 'above the Senior Wrangler', the highest-ranking woman wrangler was 'between 20 and 21'. Unfortunately for mathematics, Fawcett left academic life.

great time for work is in the Long Vacation, and that's all you'll get
when you have a teaching job.' A queer look prompted: 'You *do* work in
the Long Vacation don't you?' 'No.' 'Well, all I can say is that a young
mathematician who can't live for pleasure as much as he wants, *and* do
a proper job of work at the same time, shows a gross incompetence in
the art of life.' 'I'll think it over.' He was only a moderate (successful)
Fellowship candidate, but came on and on and became eminent.

F. S. Macaulay's F.R.S. I didn't know his subject, but Newman
(pretty omniscient) said I should get an opinion from the best expert,
Emmy Noether. I was then on the Mathematics Committee; her letter
reached me at the final election, and I had to translate slightly garbled
German at sight. All was well, and in view of his age (he taught me at St.
Paul's and was an outstanding person, with many high, and some Senior,
Wranglers to his credit) he was elected at once. In the evening I was
functioning as President (effectively permanent) of the 'Mathematical
Club' (now extinct). Macaulay came up before proceedings began to
thank me for putting him up. 'Of course I know I have no chance of
getting in.' 'You *are* in.' He blushed, retired to a corner, and glowed for
the rest of the evening. (One of my happy moments.)

We were being addressed on Green's theorem by the old curmudgeon
Joseph Larmor. I made a little preliminary speech, reproaching the
Secretaries for a date clashing with Beethoven's posthumous quartets
(duly apologized for). 'Of course, if there *were* one thing that would
reconcile me to the loss, even if Pure Mathematics is going to be put in
its place...' Arthur Berry gave an audible chuckle.

It was in fact a poor show, and Larmor didn't even understand what
I said was the 'point' of Green: to make an infinity of a function to do
positive work instead of being a disaster.

Harold Jeffreys came to consult me on mathematics. There was an
open Bible on the side table. As he went he said: 'I can't help it; I'm
dying of curiosity — that Bible.' 'The book of the film' (Samson).

From childhood I'd imagined that Samson had *pulled* the pillars,
and was puzzled how he could reach. I think also I'd seen a picture. But
in the film he used the sound rock-climbing technique of *pushing*.

I recall once saying that when I had given the same lecture several
times I couldn't help feeling that they really ought to know it by now.

Besicovitch and Harry Williams asked me what God was doing before the Creation. I said: 'Millions of words must have been written

on this; but he was doing Pure Mathematics and thought it would be a pleasant change to do some Applied.'

Harry Williams told me about 'Limbo'. A benevolent invention of St. Thomas Aquinas. Apparently you have all the natural pleasures, but may not twang the harp. Good enough for me.

The wine committee sent round a notice in 1910 (a unique event) saying that on the advice of the wine merchant they were laying down an unusually large amount of the current port (1908!) Any Fellow could come in on this and order any amount (100 dozen); the College would keep it for him in the cellar. The price would be 1/— a bottle). I couldn't be bothered.

Bertrand Russell told me that, as a highly moral and self-critical undergraduate, his reaction to Parry (The Rev. R. St. J. Parry, Tutor in Trinity) was: here is this good kind man whom everyone loves, and I just can't like him; there must be something wrong with *me*.

R. A. Leigh once asked in Hall what Hardy was like. My neighbour and I merely laughed. I should have said: 'All individuals are unique, but some are uniquer than others.'

G. H. Hardy on R. A. Herman (1917): 'The mildest of the most ferocious of Huns.' Applicable also to E. G. Gallop.

Ramanujan's Fellowship Election (1918). I am the only person who knows the facts, and they should be put on record, if only as illustrating the fantastic state of the College just after the 1914–1918 war, when there were only the unfit, and people over military age, mostly ferocious Huns. I don't see why I should suppress names.

There was much opposition. Hardy was not made an Elector, and I acted, by letter, because I was quite ill (after concussion on the top of years without proper holidays from hard work). I did get wind of the enemy's tactics from R. A. Herman, who was a close personal friend, and although he was against Ramanujan himself, he was always naïvely honest. I said: 'You can't reject an F.R.S.' '*Yes*, we thought that was a dirty trick!' I gathered also (i) that R. V. Laurence had been saying that he wasn't going to have a black man as Fellow; (ii) that 'grave doubts' were being expressed about his mental state. I met this last by getting two doctor's certificates. I spoke to E. Harrison, who was pro-Ramanujan, but not an Elector; he was obviously worried.

At the actual Election he was elected, I don't know by what majority; but it was decided that the doctor's certificates should not be read.

Matlock House
Matlock

Dear Mr Hardy,

My words are not adequate to express my thanks to you. I did not even dream of the possibility of my election. When I opened your telegram I read therein Fellow Philosophical Society instead of Royal Society — I came to know only very recently of my election to Cambridge Philosophical Society — and I was very much puzzled why you sent a telegram from Piccadilly for that. It is only after some time that I read your telegram correctly.

Please convey my heartfelt thanks to Major MacMahon and Mr Littlewood. I am sorry I didn't write to them as I did not know their addresses.

Ever yours
S. Ramanujan

Fitzroy House
16. Fitzroy Square
London
Friday.

Dear Mr Hardy,

My heartfelt thanks for your kind telegram! After you succeeded in getting me elected by the Royal Society my election at Trinity probably became very much less difficult this year.

My tooth extraction was to be done this morning, but the dentist was not able to come as he was indisposed. I have not been very well since you saw me and my temperature has been as irregular and high as was at Matlock. Dr Bolton thinks that my feverish attacks and my rheumatic pain are both due to my teeth. His idea explains very well the rheumatic pain and no other doctor gave any cause for the pain. But I can't see any connection between the teeth and the feverish attacks which reduced my body and which existed long before anything was wrong in my teeth. I shall just remove one or two teeth for the present and I do not want to trouble with the other ones now

yours ever
S. Ramanujan

Two letters from Ramanujan to Hardy.

J. M. E. Taggart, lecturing me, said I was facing an open door. But how could one be sure? And I wonder.

F. A. Simpson deprecated to Hardy my 'violence'. He replied: 'The best defensive victory is when the enemy don't move out of their trenches.'

Besicovitch's Fellowship Election (1930) In 1960 I turned out, among other old papers, my correspondence with Hardy on the Pollard-Besicovitch affair.

S. Pollard had gone mad, and opponents of Besicovitch's appointment used his diatribes, I thought unfairly. It was rejected by the casting vote of the Master, J. J. Thomson.

I was surprised by the amount of academic political work involved, and the lucidity of our correspondence, always scrawled down in pencil and copied in ink (at least on my side).

Postcard to say, later, that Besicovitch's Fellowship was assured; ASBOKQTJEL.

Wittgenstein's Fellowship Election. In the late twenties the question of making Ludwig Wittgenstein a title *B* Fellow was raised at the Council. A report from Russell seemed to make reservations, e.g. he said he was not convinced that Wittgenstein's theories were true. And any reservation is apt to be fatal to a Fellowship. Wittgenstein had not yet acquired his later status, and the Council were sceptical. They decided that *I* should have a series of talks with Wittgenstein and report, and they would act accordingly. The whole thing seems to me incredible — I was a personal friend of Wittgenstein; suppose the answer was 'no'. I said the Council were asking no light thing, but we went through with it.

Wittgenstein never finished a sentence, except to say 'It is impossible', and there was an *obligato* of 'I am amiss.' In order to understand *A* you had to hear *B*, *C* for *B*,..., *X* for *A*. I was favourably convinced after half a dozen interviews, reported, and Wittgenstein was given a *B* Fellowship.

A fee seemed hardly relevant to the fantastic picture. The Council suggested £5. C. D. Broad, saying that I had been seen staggering exhausted into Hall, got it raised to £10.

On Bertrand Russell's famous remark that the 10 commandments should be headed (like an examination paper) 'Not more than 6 to be attempted.'

I have a variant. As a Pure mathematician I can't be expected to do the *Applied* ones. Interpreting No. 1 to mean what it says in logic, I should be prepared to *do* the Pure ones. No. 2 is a sitter, No. 3 would take some initial caution, but worth it with the reward of full marks. It is good mental hygiene to take Sunday off, and I could throw in this Applied question. Nos 5 to 10 are applied. Actually my parents are dead so No. 6 is a walk-over. And I should be disappointed if I didn't get full marks on the applied 9 and 10.

The subject came to my mind one morning on a walk: on trying to recover the list I could only remember nine. I keep Authorized and Revised Bibles as works of reference, but no Prayerbooks, and was too lazy to search out a Bible. Sitting next to our Dean of Chapel, Harry Williams, I asked: 'What are the 10 commandments?' to general hilarity. The missing one was No. 2 and I developed my theory, with general sympathy!

The Trinity Council went on persecuting conscientious objectors well after the war, until we got wind of it. They got the *money*, but were deprived of the title 'Scholar'. I don't know about rights to rooms in College, but otherwise much would they care; everyone knew they were Scholars.

I was an Elector for a Professorship at Singapore (1926?). The man we elected was vetoed by the Foreign Office (Another Chamberlain) as a conscientious objector.

When Hermann Weyl became a Professor at Zürich, his class of a hundred odd dwindled to one, Mrs Weyl. He later pulled himself together and was most impressive.

When the eminent lawyer F. J. Pollock died, his family sent me two of his mathematical papers (he was an F.R.S.) for my opinion. As a lawyer he had a reputation for lucidity, but these papers were exceedingly obscure. One very long one I decided contained nothing fit for publication. The other was 'Theory of Numbers'. I struggled with it for several hours. I finally extracted something, expressible in two lines, which was good enough. I suspected it was not new, and searched Dickson's History, 3 large volumes. Sure enough the two lines were there, under Pollock's name. It is staggering to think of the amount of industry involved in the History. But Dickson did make extensive use of his pupils.

Photo of F. R. Tennant. Infinitely distinguished, and saintly. Continental intellectuals, especially Philosophers, often feel that they should be high-minded and moral beyond the ordinary. (I cannot see that a purely intellectual study of Ethics has any bearing on one's conduct). In England we tend to think that the energy spent in being saintly is a waste not to be afforded; we do an honest job of work; good enough. As a result we don't generally *look* distinguished; at any rate whatever the cause the fact is so.

Hardy (before late middle age) was another exception.

The Haldane case. J. B. S. Haldane had been a co-respondent, and the Sex Viri, a body of academic lawyers, took up the case; whether he should be deprived of his readership. I wrote a testimonial about the importance of his work, and when he proposed to resign his Fellowship Commonership at Trinity I advised him to *offer* to resign. To the fury of the Puritans on the Council the rest of us turned the offer down.

The Sex Viri were 4 to 2 for sacking him. Lawyers are reputed to be able to drive a coach and horses through most regulations, and one would have expected the Sex Viri (or its majority) to go ahead and sack. Instead they wrote to the Chancellor saying that they didn't know whether they had to be unanimous or not! He took Council's opinion. One would have thought that in a doubtful legal question the Council would decide, in the existing state of opinion in the outside world, in favour of Haldane, that unanimity was necessary. He was told that it was not necessary.

There was then a comic episode. The outstanding two were Bond, Master of Trinity Hall (and a close friend of mine), and X. Bond was the most high-minded of Liberals and always acted according to his conscience, and damn the consequences. After the Council's verdict, Bond announced that he had changed his mind! Now X was a man who could not bear ever to be in a minority of one, and he also changed his mind. So the Sex Viri *were* unanimous.

Haldane appealed. The Court appointed by the Council of the Senate seemed packed against Haldane. The chairman was a lawyer notorious for his savage majority. There was also an M. P. for the University, with his constituency mostly country parsons. (However he wrote to me about the case and I think he favoured Haldane.)

Now when it came to the point it emerged that the Sex Viri had done something fantastically improper. (I understood that they had not heard Haldane's defence, but have been told that it wasn't quite that, though almost as bad.) The Court immediately reversed the Sex Viri's decision and Haldane survived.

It was only the giggles of the outside world that suggested there was anything odd about the Sex Viri. It has been — pusillanimously — changed to Septem Viri, on the ostensible ground that that ensured a majority one way.

J. Proudman, ex Fellow of Trinity, was Professor of Mathematics at Liverpool. He was supposed also to be responsible for the Astronomical Telescope, but ignored it. N. R. Campbell, a Classical lecturer at Liverpool, told the following story. He had an amateur interest in Astronomy, and was constantly asking Proudman to be allowed to use the Telescope. One evening they did go. Proudman said: 'There are no planets visible.' Campbell took the assistant outside and pointed at Jupiter; 'What's that?' So they agreed that Campbell should have his look. But the rope controlling the telescope wouldn't move it; finally it broke and Proudman was injured.

D. A. Winstanley on X, a tutor. His room makes one think that there should be a notice: 'Please adjust your dress before leaving.'

My first visit to the University Library (1945). I wanted the last two figures of the date of a book. I had owned the book, but a pupil must have failed to return it. I wrote to the author, to find that the book was by another man of the same name, who was inaccessible. Hardy, who had it, was away. So I went to the University Library. I had 'read' for them for years and the book was one of a series for which there was a standing order. It was not to be found in the catalogue, either under the series or the author's name. So finally I got into Hardy's rooms.

I suppose a historian spends about half his time in such activities.

The Lancashire dialect reverses the 'u' sounds in 'put' and 'putt'. A Lancastrian gave a lecture to a large mixed class on the Antarctic expedition led by Sir Vivian Fuchs. He called him Fucks throughout. A colleague remonstrated after the lecture: 'You know his name is Fuchs.' 'Oh yes; but I didn't like to say it with all those ladies present.'

I told this story in the Combination Room on a Saturday night; there was a Chaplain opposite. On the following Sunday, Leon Radzinowicz, later Sir Leon, came up as we were leaving: 'I've been sitting between two Bishops, and they told me Littlewood's story.' (Via the Chaplain.)

Montagu Butler on Fellowship Elections. 'It is often impossible to decide between the last two candidates.' 'What do you do?' 'I try to make up my mind whom I think to be the really best man.' (Told me as authentic.)

A. E. Housman's speech at the dinner held at University College, London, Kennedy Professor: 'Cambridge University has seen many strange sights; Wordsworth drunk, and Porson sober; and here stand I, a better scholar than Wordsworth, a better poet than Porson, betwixt and between.'

The Listener, 16.4.59, perpetrated the following version, which repays close study.

'The Hall of Trinity has seen many unusual sights. It has seen Porson sober. It has seen Wordsworth drunk. And I stand here, a better scholar than one and a better poet than the other, betwixt and between.'

New Problems. Complete indecision between yes and no in an exciting new problem is agonizing. When you all go out one way (either yes or no), the thought keeps nagging that it's an even chance that it ought to be the other way.

The difference when you *do* know (when, for example, we are looking for a new proof) is enormous. Like a Bridge problem 'if the thing *is* possible it must be necessary to lead the Ace and trump with the Ace in Dummy.'

When descending the West Chimney at Kern Knott's in wet and bad light, at one point my first reaction was 'Impossible', then at once, 'There *must* be a hidden hold *here*', and put my hand on it.

Mathematics and engineering. W. M. Fletcher (tutor) conceived the idea (*c*.1912) that Engineering students should be taught some 'real' mathematics by the *mathematical* staff — 'contact with great minds'. The hardworked, hardboiled, and lazy devils hated it as much as I did, to whom, as junior, all dirty work then fell. I asked F. J. Dykes (then sole Lecturer in Engineering) what he would like me to select; all he said was: 'Give the buggers plenty of slide-rule.'

A record Ph.D. At Cambridge a candidate is expected to have adequate knowledge of the literature directly connected with his thesis, and a reasonable background of more general work. (Not at London, or Oxford. With a thesis about integral functions of zero order, complete ignorance of the theory of finite order was held to be irrelevant.) I suggested (*c*.1923) a subject to X (one of about 10 research pupils), saying that as far as I know (I was frankly not an expert) the only relevant literature was a paper by Jacobi; I didn't think it *was* going to be relevant, but he was to read it. The thesis on its merits was more than adequate. My co-examiner F. P. Ramsey, who concealed behind his (then) contemptuous and 'King's' highbrow exterior a high standard of conscientiousness for himself and others, read Jacobi to the end, and found that all the important part of the thesis was there. Up to the point I had read, it looked very much as if there was nothing doing. X had read to this point, come to the same conclusion, *and read no further*. The paper does later take a quite unexpected turn. Ramsey, utterly shocked, obviously thought the case should be turned down. My view was that, while X's slackness was scandalous and should receive heavy censure, we should award the Ph.D. for brains and promise (it was obvious that there was no deliberate fraud), and I was reporting 'yes', though with all the cards on the table. I told Ramsey I didn't expect others to necessarily agree with me, 'the Government Whips are off', and I said he should say 'no' if that was what he felt. He didn't. The next hurdle was the Degree Committee. I enlarged verbally on the case, and again said the Whips were off. Perhaps they had all had good lunches, but there was not only a majority vote 'for', but, what I would not have expected, unanimity. The Board of Research Studies, of course, would have liked nothing better than to send the case back for reconsideration, but faced with unanimity they didn't dare. So X is a Ph.D. (and on the whole has fully justified it).

The Astronomer's fallacy. It is very hard to make a *random* selection of stars. If, for example, you see a star (with the naked eye) it is probably bright (as stars go).

A lecturer was once making the point that middle class families were smaller than lower class ones. As a test he asked everyone to write down the number of children in his family. The average was larger than the lower class average.

The obvious point he overlooked were that zero families were unrepresented in the audience. But further, families of n have a probability of being represented proportional to n; with all this, the result is to be expected.

Swerve of a cricket ball. We were always hearing about the effects of the heavy damp atmosphere of Manchester, and so on.

Experiments in Australia (instigated by R. A. Lyttleton) elicited very unexpected results. There is, for example, an optimum speed for swerve; Trueman swerves, Tyson doesn't, and was better with the new ball (he was *too* fast); the optimum is round about Bedser. The final general conclusion was that in future you believe what cricketers say, and humbly design experiments accordingly. (On the other hand, they are completely hopeless about the 'law of averages'.)

Speed in creative mathematics. I say there is a speed over 10–20 seconds, 10–20 minutes, 1–2 hours, 1–2 days, weeks, months, years, decades. *And people can be in and out in different classes.*

With a collection of really difficult problems, nothing happens in a year; much happens in 10 years. (This leaves the beginner cold.)

A. E. Ingham's yearly pace is snail-like; over 20 years he is most impressive.

This question came up over Philip Hall's election. He had published nothing for 11 years. (Partly because of his time at Bletchley during the war.)

At a guess I should say I was (1960) quick over 20 minutes, 6 weeks, 1 year.

Competence, or high competence, and high creative ability. No correlation, of course.

Most of the best work starts in hopeless muddle and floundering, sustained on the 'smell' that something is there.

This is shown notoriously, by Beethoven's note-books. As for competence, the fifth- (tenth?)-rate Saint-Saëns is said, after glancing at the score of Tristan while waiting for Wagner to appear, to have sat down at the piano and played it to him (to his ecstasy).

For that matter, Beethoven could improvise brilliantly. There is a story that to entertain company, he took a viola part of a score, put it upside down on the piano rack, and on this as a given base improvised a long and brilliant Passacaglia.

It is possible for a mathematician to be 'too strong' for a given occasion. He forces through, where another might be driven to a different, and possibly more fruitful, approach. (So a rock climber might force a dreadful crack, instead of finding a subtle and delicate route.)

The 'undergrowth'. In a new subject (or new to oneself) there can be a long preliminary process of getting to the essential core. At *this* process a first-rate mathematician is little, if at all, better than a Ph.D. student. The latter is bad at understanding this: *he* has cleared away the undergrowth in the course of 3 weeks, and he expects his supervisor to get the point in 5 minutes.

I once got a manuscript on a suggested problem from a man of Fellowship class. It was quite good, but it stopped short of the suggested positive result. It had an essential idea breaking the back of the problem, and I judged that the remaining cases would almost certainly go, but there was no obvious proof. The author ultimately explained that he thought it would be an insult to trouble me with the missing proof.

Queer ways that theorems get proved.

(1) *Bloch's theorem.* One of the queerest things in mathematics, and one might judge that only a madman could do it. He was aiming at an 'elementary' proof of Picard's theorem, an impudently 'damn fool idea'. With this as a start it is just a reasonable stroke of insight to *conjecture* Bloch's theorem. The result once conjectured (and being true), a proof was, of course, bound to emerge sooner or later. But, to keep up the air of farce to the end, the proof itself is crazy.

(The day after I had lectured on Bloch's theorem, I got an offprint from Turán with a new and attractive proof, and I presented it to a distinguished class next lecture. By the next lecture again I had realized that the proof was fallacious. It had got past all the class, including H. D. Ursell. 15 years later, Turán wrote to me about other things, and I think he must have mentioned Bloch. At any rate I broke the news (it *was* news) to him.)

(2) There is a fine example concerning Marcel Riesz which I recount later on in *The Mathematician's Art of Work*, p. 193.

*(3) If $a_n \geq 0$, $\sum a_n = 1$, $C = \sum a_n(1 - \cos n\theta)$ *and* $S = \sum a_n \sin n\theta$ then $\int \frac{\theta^\alpha}{1-C+|S|}$ *converges as* $\theta \to +0$ *for* $\alpha > 0$. This was sent by A. S. Besicovitch from the U.S.A. and was a problem raised by a statistician. (It was very baffling and worrying.)*

I got a certain distance, reducing the result (in the first place with $\alpha > \frac{1}{2}$) to an unproved Lemma. I had what looked like a promising idea for this, but it was fallacious. In the middle of a three week holiday — mathematics completely below the horizon — the idea came again out of the blue when I was in bed. I *forgot* it had been rejected as fallacious, and this forgetting did the trick; because *this* time I noticed that it did prove *something*, and, by what was nearly, or quite, automatic writing, a proof got written down which deduced the Lemma from the something. (I did once, when about 21, experience automatic writing in fairly easy exploring calculations — the pen acted, and my mind was elsewhere.)

(4) Shortly after recording the above, I did experience automatic writing again.* When working on the $M_1 < (1 - c)M_2$ problem for real

trigonometrical polynomials my pencil wrote down the formula

$$\int_0^{2\pi} \left| f_n(\theta) - f_n(\theta + \tfrac{\alpha\pi}{n}) \right|^2 d\theta = \pi a n, \quad a = 1 - \frac{\sin \alpha\pi}{\alpha\pi}$$

for no reason at all, and almost unattended by consciousness.* On the face of it the formula has no apparent connexion with the problem, but it turns out to be an essential key to the proof. (Incidentally my first arguments were quite fallacious, but I was inevitably on the right track.)

This inspiration came at the end of a painful period in Davos. First mild flu, not even going to bed; but the effects lasted for three months. But worse, there was a practically unbroken four weeks of Föhn (the worst season in living memory). On looking back, I really think my brain wouldn't function properly (Föhn affects me as it does the Swiss, who say they can neither think nor use their muscles, and masseurs have to work overtime). For one day the Föhn stopped; I had a flow of ideas towards the end of my (1,000 feet) walk up to Schatzalp, and in the restaurant I did the automatic writing.

If we may reject divine bounty, it happened exactly as if my sub-consciousness knew the thing all the time, and finally got impatient.

The Trinity Entrance Examination.

Hardy (c.1904) was woken at 7 a.m. by a man from the University Press to say that the paper for 9.00 that morning had not been received. He broke into the library and selected questions at random from previous years. The typesetting and proof-reading were done by the Press, Hardy duly catching the 8.45 as he had planned. There were many misprints (so the story goes, though I don't see why) and the paper was doctored for the record.

Before 1914, I was Junior on the staff (of four) and did all the dirty work. After 1918, I was Senior, but we became democratic and shared the dirty work. Except that I had to examine regularly for Fellowships, and in exchange no longer did the Entrance.

In one paper (semi-optional) it was a game for the Examiner to finish marking all the scripts before the 3 hours were up. One was apt to be defeated by one or two people who could do something and stayed to the end.

Before 1905, the College was full (it had a fixed number for under-graduates, and still has much the same one), but only by taking any riff-raff. Some energetic Tutors worked up connexions with Headmasters etc., and it became possible to reject 10% on the examination. (The change was startling. New Court became a place of gentlemanly and civilized behaviour, instead of a beer-garden, with drunken noise from noon onwards.)

There were three categories — rejects outright by the examiners, acceptances outright, and 'doubtful'. In the last the Tutors picked and chose (every now and then by dubious methods perhaps, but life would be impossible otherwise) and what they liked was for the doubtful class to be as large as possible.

My only interesting experience was to lose C. N. Lowe for the College. Not in the least by *failing* him. But the Tutors asked, with mysterious emotion as it seemed to me (who had not heard of him), that he should be given an Exhibition. This was so preposterous that having to refuse didn't embarrass me (it might have been possible to compose special questions with due notice). He would have come for an Exhibition, but, having doubtless an offer of a Major Scholarship at the Hall in his pocket, he went there.

About 1912 I met a man in Bideford, who said 'I hear you are a Trinity man, you must be awfully brainy.' He didn't know I was a Fellow (nor, probably indeed, what such a thing was); it was merely the notorious severity of the Entrance examination. (Lord Randolph Churchill would have liked to go to Balliol, but the Entrance Examination was too severe.)

'Gentleman, The Boar's Head.' Speech at the Christmas feast at an Oxford College. The boar's head used throughout as the symbol for the rude vigour of our ancestors, contrasted with the namby-pamby substitutes of today. The speaker, however, was the Rev. W. A. Spooner, Warden of New College, Oxford.

Sermon by Montagu Butler. 'Shall we wake with the wise virgins, or sleep with the foolish virgins?'

G. H. Hardy once scrawled a five-letter schoolboy term of contempt on a pupil's example. On second thoughts tried to erase it, in vain. He told the pupil the paper was lost. The pupil happened to be hyperconscientious, and also very pious, and kept clamouring for its return.

Vyvyan told me recently that he had a similar experience.

Cambridge Common Rooms are full of people talking about the decadence of Oxford; and Oxford ones are full of people talking about the decadence of — Oxford. (*c.*1930. Clearly from Oxford, and incidentally the first clause at least was quite false.)

(In 1947, over the port, I spotted that my neighbourhood guest was an Oxford man: 'It certainly wasn't your manner; must be an inferiority complex on my part.' He took it for complete earnest, and tried to cheer me up.)

Mrs Rouse Ball. My husband always says, on a cold day there's nothing like a fire.

Mrs Hammond, showing H. A. Hollond and me round the Trumpington cemetery at someone's funeral. 'Here is Aldis Wright, and here is Mrs Rouse Ball next door: I do think that is so nice.' To our surprised reaction: 'Trinity, you know.'

Generosity. In pre-1914 days, Henry Jackson offered to show a party of Americans round Trinity, and did everything, including the things possible only to Fellows. When they parted they tipped him 4d.

C. D. Ellis on X, grandson of a famous and son of a prominent man. 'It is a fair extrapolation to the third generation (*sotto voce*: the fact is the poor boy is practically feeble-minded).' (Aimed at the man in the middle.)

Improvement. At the Royal Society Council an application for a grant for a Spitzbergen expedition praised the varied career of the proposed leader: Harrow, Officer in the Guards (First World War), business job, enlistment in the Merchant Service as ordinary seaman, Oxford Undergraduate. I commented: 'At least we can say he has always been on the upgrade.'

Never resign; your idea that they will give way is mistaken.

J. J. Sylvester sent a paper to the London Mathematical Society. His covering letter explained, as usual, that this was the most important result in the subject for 20 years. The Secretary replied that he agreed entirely with Sylvester's opinion of the paper; but Sylvester had actually published the result in the L.M.S. five years before.

Bargain. When Trinity once (before 1914) had some wine at a lowish price, but not satisfactory and also not selling, they doubled the price, whereupon the idle rich bought it all up.

T. J. I'A. Bromwich once in the 30s, wanted my address. The Porters' Lodge reply, by still standing order, was 'Letters will be forwarded'. In the old days you did not enquire where Fellows had been in the Vacation.

A. A. Markov. (via Besicovitch). A Ph.D. this having admittedly failed, the other examiners were in favour of leaving it at that. Markov wished to read the man a severe lecture on the enormity of his performance, but allowed himself to be overruled. On his death-bed he said he had never forgiven himself for this weakness, and it saddened his end.

Difficult collaboration. A. S. Besicovitch had moderate difficulties with others, but was driven crazy collaborating with Harald Bohr (as was I). Bohr had a vivid sense of humour and quite understood, and told the Danish story of the two executions. In the Middle Ages town *A* had plain executions, in *B* they had the last extremity of horror. To a woman whose son had been executed: we are all *so* pleased, what luck it was in *A*.

Bertrand Russell had a story that when he was at *X* everyone began: Pleased to meet you. The University of *X* has in the last 6 months suffered three irreparable losses; Professor *A*, etc. Then he determined to get in first, and rushed out the complete formula. Nevertheless, '*As* you say, the University...'

When you make your speech at a Trinity Fellowship Election, do not expect them to break into irrepressible applause; no one will blink an eyelid. They have taken it in none the less.

Creative workers need drink at night, 'Roses and dung'. (Or: mathematicians read 'rubbish'.) An experimentalist, having spent the day looking for the leak, has had a perfect mental rest by dinner time, and overflows with minor mental activity.

A. N. Whitehead once got a man a Trinity Scholarship on the ground that he was such a nice muddle-headed fellow.

Spooner. At a Fellowship Election at New College, there were four Electors *plus* Spooner. Spooner said one Elector was in favour of *X* (whose claims were *prima facie* on the snobbish side); the other three were against. Since Spooner's vote as Warden counted 2, this made a tie. Spooner then gave his casting vote to *X*.

Henry Sidgwick at a Fellowship Election: 'I can see this is bosh, of course; but t–tell me; is it the right k–k–kind of bosh?'

Copyright. It is legally impossible to copyright a *theory* — or anything that published true universal[1] facts and nothing else. (A new

---

[1]Bradshaw *is* copyright.

Ballistic *Theory* was, I gathered, not secret in the First World War.) You can, however, copyright a set of mathematical tables provided they contain mistakes, and Comrie made a practice of introducing them in his collections.

The Passionate Pilgrim. There is a copy in the Trinity Library. Presented by an eighteenth century Fellow and accompanied by a letter apologizing for the poor condition and the fact that it cost him only $1\frac{1}{2}$d. There is only one other copy, slightly inferior in condition, and it was last sold for about £7,000.

The Cuckoo. To decide a bet Ann wrote from Cornwall (August 1949) for the real facts. I consulted all the Fellows from Trinity I came across (it was deep Vacation) with no result, though everyone knew it was an interesting problem. Then I ran into Adams (Librarian); no result. But I said I had a vague idea there was a book by a man with a name like 'Chanee'. This being about *books*, not about *things*, I got an instant answer. 'Oh yes, Chanee is an old Trinity man; he presented us with a copy; I will have it sent to your rooms.'

The Library. On the fact of it it would be easy to steal from the priceless show boxes in the Library. Assistants are hardly to be seen, and there would be a telephone call at the critical moment.

The only market would doubtless be American millionaires with 'private' collections — no questions asked (Göring in the '30s).

Prediction. In 1908 Herman quoted to my father an extract from the main report on my *first* Fellowship thesis. 'He is to be a great mathematician, and his work is singularly mature.'

(Herman, on the evidence of my Minor Scholarship and perhaps a first term's work, broke it to my father that he couldn't expect much of me. I'm not sure when this was reversed, and my father knew the ropes enough not to build too much on a Senior Wranglership. I remember, to my surprise, as being a prodigy of 14 in South Africa, his implying that I might expect at Cambridge to be nearly as good as himself, or possibly a shade better. The Fellowship report was something of a surprise; he said 'I feel like a hen that has hatched an eagle.')

R. V. Laurence was making trouble about mathemtical teaching. I asked H. A. Hollond, 'Is he trying to drive me away?' Hollond said 'I think perhaps yes.' Hardy, back as Professor, went to see Laurence; after all they were old common members of a Port Club in their youth. He told me afterwards what he said. 'If Manchester thought there was any hope they would send seven letters on the off-chance that he might fall for the last one.'

According to Besicovitch I once remarked that for a Fellowship dissertation to succeed it ought to be something that would take me at least 3 weeks. I said 'I was young, but it was a most arrogant remark.' Besicovitch 'Not in the least.' Hardy once repeated the remark to H. Bateman, who shouted with laughter; only a rare dissertation would qualify.

When A. Zygmund was in England in 1926 I wondered if he was avoiding me for some reason. I then heard from Besicovitch that he said it was not right to come to me because I spent too much of my time on pupils, and it ought to be kept for more important things. This thrilled me at the time, compliments to my character being then nonexistent. Now that I *have* (according to Besicovitch) become a half-saint (in academic matters), for want of better things to do, I am no longer interested in my moral character or what people think of it. (This is about the one compensation for getting old.)

ODDS AND ENDS

On being given a bottle of whisky by his boss a man said; 'It was just right; if it had been better he wouldn't have given it me; if it was worse I couldn't have drunk it.'

Some Arts, e.g. Opera, are 'impure'. The most impure subject is New Testament criticism; you have to weigh such questions as whether the Son of God could make a mistake in grammar.

If we must have Strauss let it be Johann; if we must have Richard let it be Wagner.

Perfect greed casteth out fear.

D. H. Robertson remarked over the hot charcoal biscuits in the Combination Room (1947): 'They are burning coffee in Brazil, and we are eating coal in England.'

J. P. Postgate. ('That word occurs three times in the known literature.') In Hall he was maintaining that the Roman Empire had to all intents everything we had. I asked whether they had asparagus (which we were eating). 'Oh yes.' 'What was it called?' 'Asparagus.'

Einstein in a lecture said: 'This has been done elegantly by Minkowski; but chalk is cheaper than grey matter, and we will do it as it comes.' (Via Pólya, who was there.)

A Trinity Fellow from Ceylon asked me how seriously Ramanujan took his religious observances as to food, etc. 'He was dressing for dinner in the jungle.'

The Solera principle from Madeira is that when you get something unusually good you fill up when it is a quarter full with any old stuff.

There are Solera Madeiras more than 100 years old. We were discussing this in the Combination Room, and I had a Bishop opposite. He was very nice, we had had two glasses of port, and I felt I could risk it. 'It is the principle for making Bishops.'

He makes difficult things difficult.

X knows the right thing to say about the label on a wine-bottle, but didn't know a corked bottle when he met it. The image of the English Tripos.

War time adaptation of a Cornish prayer about wrecks. 'We do not pray, Oh Lord, that Fellows should be killed by a bomb; but if, Oh Lord, Trinity Fellows should be killed, we pray, Oh Lord, that they may be *X* and *Y*.'

X can extract a superiority complex from any raw material whatsoever.

Oxford tutors waste much time in turning good and genuine thirds into spurious seconds.

When asked what I did between spells of work in Cornwall I answered, truthfully, 'Household chores.' Everyone giggled.

X has every gift but genius.

'Nobody could *be* so distinguished as he (Eric Walker) *looks*.

'Let us talk about something less important and less boring than politics'.

On knowing one's place. Most people do within 20%. My life is not poisoned because I haven't got a George Cross, and am not the most handsome and well-dressed man in Trinity. (When I said this to C. W. Oatley he wickedly said: 'Much the most handsome.')

But the lives of 'failed F.R.S.'s' *are* poisoned. They don't realise the height of today's standard, which is exactly poised to maximize the excess of pain over pleasure.

Harmer was complaining that they had elected a Professor of language *X*, who was unable to speak it or understand it when spoken. 'Glad to see the electors don't let themselves be deterred by trivial deficiencies from electing the obviously right man.'

E. Harrison: 'Is it true that philosophy has never proved that something exists?' Bertrand Russell: 'Yes, and the evidence for it is purely empirical.'

'I don't like him, but I don't attribute my personal tastes to the Almighty.'

We were exhorted during the war to have only 5 inch baths, like the King. 'All very well, but the King is only 5 inches thick, what about Kitson Clark?' (G. S. R. Kitson Clark doesn't like references to his figure, but I couldn't resist it.)

One should not jest about serious matters, such as the 'theory of mathematical characters', but, as in the novels of H. G. Wells, the 'principal character' is always 'improper'.

Thackeray, on being warned that neat brandy ruined the coats of the stomach : 'Well, the old boy will just have to get along in shirt-sleeves.'

I distinctly remember saying at High Table that 'Sherlock Holmes has never been the same since he fell over that cliff.' This is quoted verbatim (with its technical inaccuracy) by Conan Doyle in a Preface, as said 'by a witty critic'. Coincidence? Or do remarks speak more than you'd think?

'Was he a careful writer?' 'Well, he cabled a semi-colon from New York.' (My pupils won't bother to punctuate properly.)

Viola tone. This admirable element in a quartet or orchestra arises from the fact that as the viola is played under the chin, it has to be shallow for its size. The possibilities of new instruments today would seem almost unlimited: any blend of tones.

Metcalfe (Manciple at one time). In a period when people got drunk at May-week Balls, a Jesus boxing blue clamoured for champagne after the 3 a.m. limit. Metcalfe was called in, and the man attacked him. Metcalfe knocked him out. Everyone afterwards behaved admirably — the Blue said he'd deserved it, and he'd never met a better straight left. Metcalfe had to be admonished by the Junior Bursar, no doubt with a twinkle in his eye. Metcalfe didn't like congratulations!

There is a popular saying that when you sneeze you are very near death. No doubt for a small fraction of a second one is in a state which would be be fatal if it continued. I always wondered how such a sophisticated idea could get into folk-lore. I now find that it dates from the Black Death. Sneezing was the first symptom.

People have died from hiccupping, but never from sneezing. I remember being asked, aged 12, by a schoolmaster, how to define a hiccup, and that (to his obvious astonishment) I answered instantly: 'An involuntary contraction of the diaphragm.' I *had* been reading a simple book about the human body, but it must have been partly a lucky inspiration.

This was Wynberg, near Cape Town. I was precocious and odd, but tolerated as a freak even by tough Afrikaaners.

A girl cousin of mine had very bad astigmatism. In the early 1900's correction of vision was very casual and you often got merely the nearest they could do with a convex or a concave lens. Her father was a doctor and it occurred to him to look into her case. With her glasses on she exclaimed 'I can see the cow's tail.' She promptly became rather a good artist.

I had a less extreme experience. I'd never worn glasses. One night, accompanying my doctor uncle on an evening visit, I said 'Why does one see half a dozen crescent moons?' This showed I was astigmatic ($2\frac{1}{2}$ dioptres). When this was corrected I could see, for the first time, the detailed foliage of a tree. No wonder I had been a poor field at cricket. I find it hard to reconcile this with the fact that as a child I could not only meet the then fashionable challenge to write the Lord's Prayer within the size of a threepenny-bit, but could write it *twice*.

At one time in Cornwall we had two dogs, one, L, a whippet type and the other, C, a spaniel type, very intelligent (*he* adopted *us* and we had to square his first master). We used to go down and up a fairly stiff 'moderate climb' to bathe. One day the dogs began to follow us, *very* slowly and cautiously. They gradually got to know every move, and ultimately, with a sort of controlled fall, used to descend the 180 feet in half a minute. Coming up there was one problem very difficult for a dog. L found out how to do a delicate traverse (he once fell off and broke a leg). C, more clumsy, used to force up a small gully. He sometimes had to make 6 or more shots at it. He couldn't bear being lifted up, whereas L, on occasions when he wasn't doing his traverse, *loved* it.

Once L, following behind us, did a jump up to a ledge too carelessly. His forepaws were just over the ledge and he did a wonderful recovery by inching up with his hind ones. We applauded. He gave a sheepish wag of the tail, knowing we meant he was a damned fool but applauded his recovery.

C once did a most intelligent thing. L had got into trouble. Instead of coming up to us to try to explain he went down to the bottom and barked steadily, which we understood.

C was devoted to Ann, as only a dog can be. He used to be most worried when she swam far out.

The button-shooter. Sitting after dinner at Arosa, we had opposite a plump man in a very elegant evening dress. Presently there was a loud ping, and F. W. Aston's sister jumped at being hit in the chest. It was a pearl button shot from the man's waistcoat. We all dissolved in laughter.

I was once (after concussion on top of overwork) unable to read. It is a revelation to find out how much of life is reading. I tried knitting, but without enough experience found it worse than reading.

The blind Fellow of the College, H. M. Taylor, had a boy to read the papers to him, and ignored the frightful Cambridge accent.

A Frenchman was committed to the Bastille for 20 years from 1788, and forgotten. He spent the time writing a eulogy of Louis XVI. When released they explained to him: 'Well it's rather awkward; rather a lot has happened in the meantime.' 'Quoi, il est mort?' (via Russell)

Early recollections of mine

I attended the school gymnastics display, long after my bed-time, and a great thrill. Finally I was hoisted up and hung alone on the bar. I was quite aware of the exact nature of the thunders of applause, but I loved it.

On my sixth birthday, I was told I was now a moral being, and responsible for my actions. As I had agonized over my sins since an infinite past, it was a bitter blow to learn that all that had been needless.

Aged six, I was told, on asking a question, that I was too young to understand. My reaction was: what nonsense, I can understand anything. Children arrive at the adult way of thinking of space, quantity, etc., at different ages, but I'm sure I was 'adult' at six.

In 1899, when I was about 13, and so a possible chaperon, I escorted the daughter of the Commissioner for Education to a speech by Rhodes. 'Kruger will climb down without firing a shot.' I didn't believe it, and by association felt sure the First World War would begin disastrously.

At about the same time, I tried the experiment of drinking nothing from the time of getting up. After lunch, which was hard work, I cycled 8 miles through dust and a temperature in the 90's. By 4.30 I was experiencing the full horrors of thirst, which practically no-one ever gets.

Since then 'thirst' has never troubled me (e.g. on mountains): I can always survive till evening without discomfort.

Cattery, just or unjust

W. Whewell, Master of Trinity, after the inaugural lecture of the new Professor of History. 'Who would have thought we should so soon regret poor Kingsley?'

X has a career of the highest promise behind him.

Baldwin always hits the nail on the head; but the nail doesn't go in. (Witty only if true, but it was certainly what I used to think.)

A father, in the last hope of making his twin sons distinguishable, sent them to Schools X and Y. The one became X's most perfect type of gentleman, the other the most offensive bounder ever known at Y. But still...
(This is extremely liable to be told the wrong way round.)

Compressing the maximum number of words into the minimum of meaning. (Churchill on Ramsey MacDonald.)

'X is ambisinister'. (J. E. L. on E. A. Milne. Unfortunately antici-pated, I learn, by Aristophanes.)

Housman used to keep a notebook of witty and nasty remarks, awaiting opportunity.

Wise-cracks, gags, etc.

All the sensible and enlightened people said one thing, and all the damned fools said the other; and the damned fools were right. (Outbreak of 1914 war.)

A statesman is a politician held upright by pressure from all sides.

The measure of our intellectual capacity is the capacity to feel less and less satisfied with our answers to better and better problems. (C. W. Churchman.)

Montagu Butler explained that his speech did not mean what it meant.

G. H. Hardy finished a review of Bieberbach on the two kinds of mathematics. Bieberbach, a Jew, more Nazi than the Nazis: '...may well have to exercise discretion, but in this case one is forced to the uncharitable conclusion that he really means what he says.'

A. S. Besicovitch's comment on this: 'It is much more often than you would think that people mean what they say.'

The extreme materialist Philosophy: Der Mensch ist was er isst. ('isst' = eats.)

Charm is an attribute only of those who don't know they've got it. (Alistair Cooke.)

He that bloweth not his own trumpet, his trumpet shall not be blown.

Nothing is more unpleasant than a virtuous person with a mean mind. A highly developed moral nature joined to an undeveloped intellectual nature, an undeveloped artistic nature, is of necessity repulsive. (William Bagehot.)

Stories

In my second term, the Junior Bursar (J. W. Capstick) summoned me because my cheque for the College Bill had crashed. (My allowance from my Father had not arrived in time, but at the actual moment all was well.) Half my mind dwelt on prison, etc. All Capstick said was: 'There are several, what is your name?'

The examination personality. After a certain amount of them, and when they become severe, there is a definite personality. One never feels the strain at the time, and concentration throughout each three hour paper is automatic. One does not have to make a conscious *effort*.

(I knew already, from experience at St Paul's, that if one arrived with bad toothache, it would vanish for the occasion.) The final examination, with the old Mathematical Tripos as the worst case, could be very great. In my case it left me emotionally drained; I cared not a pin for anything, and this lasted about three days. (Part II, Old, was not nearly so bad; and anyway, I felt I was comfortably on top of it.)

Those in the running for the Senior Wrangler did not go to the Senate House to hear the results read. From my room in the Hostel I heard 9 o'clock strike, and it seemed that in no time a party of my friends arrived with the news.

Forsyth got the news from his bedmaker: 'We've done it, Sir. He was a fair certainty.'

I first met Miss Cartwright as an Oxford Ph.D. candidate. At the Oral my co-examiner asked a question so silly and unreal that she was completely taken aback, and blushed. I was able to get in a wink, and I think it restored her nerve.

In 1935, in the middle of the Lent term, I suddenly realized I was very stale, and how wonderful it would be to spend three or four days at Treen. I left by a fast train allowing comfortable time to catch the 10.30 express, and finished my breakfast coffee out of a 3d Thermos which I left in the rack. At Paddington the main ticket office was for the moment *closed.* They explained: 'We don't have very much traffic at this time of the year.' I had a carriage to myself, with a pot of China tea brought to me at 11.00, an excellent and comfortable lunch and a pot of China tea at about 4.15. I arrived in Penzance at about 4.40, and got to Treen in time for a bathe — the weather stayed fine and windless. Having left Thursday and returned Tuesday (so had to report one week with more than the allowance of two days' absence), I had over four days complete change and rest. On return it seemed like a fortnight, not to say month. Those were the days.

An American with a diplomatic job in Russia after the War, had a mistress with whom he lived very happily. But in a quarrel he burst out: 'You're nothing but a third-rate whore.' 'In the U.S.S.R. nobody is third-rate.'

The greyhound. It was giving two small dogs a good time by being chased. It made enormous leaps, but came down in practically the same place, so they could just keep up.

I was at the quarter centenary of Liverpool University (living for days almost entirely in the company of 20 Vice-Chancellors). There was an evening show in an immense hall. Arrows for 'Ladies' and 'Gentlemen' pointed in opposite directions. But after walking for about 100 yards a companion and I found that (because of some temporary building) the final destinations were together and separated only be a flimsy partition. 'This is an image of life.'

We once elected a Fellowship candidate on a dissertation that was an immense inverted pyramid poised on a point that was not there. Nothing was to be rescued. The facts were all put before the meeting, and there was agreement that the 'promise' was good enough. It was, however, a remarkably extreme case of the principle. In fact the election proved justified, and the man is now an F.R.S.

(The thesis was not strictly my affair, but I had thought, the previous year, that it was a bit too good to be true, and that time I made a strenuous effort and ran the fallacy to earth. The official referee, highly eminent, was an odd man.)

Soon after I was Senior Wrangler I was at a Garden Party in Bideford. As you go West life improves in most ways, but intellect is an exception, and I was very much a Lion. It happened that the Berrys also were (by extraordinary coincidence) at the Party. I ran into Arthur Berry, and he took occasion to remark, 'I remember, when *I* was Senior Wrangler...' He was the kindest of men, but he told me years later that he just couldn't resist it. (I didn't know at the time he had been one.)

In 1912 an insane Fellow (P. J. Pearce) was thought to have escaped from his asylum and likely to come to a College meeting with a revolver. The doorkeepers chosen for toughness were H. F. Stewart and J. E. Littlewood with A. V. Hill runner up.

In 1941, the Fellows had Firefighting training in small batches: my lot contained G. M. Hardy, G. M. Trevelyan (Master), F. A. Simpson, and someone tough; but I was unanimously assumed to be the toughest, to do the main job.

I once ignored the various coloured letters asking for a return of Income. When they had used the last one (mentioning possible imprisonment) they paused, then began all over again with a letter in an ordinary envelope. 'We quite understand how difficult it is for great minds like yours to ... etc.; but the situation *is* rather awkward. If you would be so kind as to call on us sometime we would fill up your form for you.' I was, of course, so touched that I sent a return by the next post.

When engaged on a tough bit of war-work (about '43) I wrote explaining that I didn't want to interrupt it, but would send a return when it was finished.

Sometimes one feels an instinctive urge not to interrupt work with something needing close attention. It is wiser to heed this (within reason; it mustn't become an obsession), and indeed *all* instinctive urges.

G. H. Hardy said he thought on paper ('with my pen'). He wrote everything out (in his invariably admirable handwriting), scrapping and copying whenever a page got into a mess. When I am thinking about a difficult problem everything goes onto a single page — all over the place with odd equations, diagrams, rings. However appalling the mess, I feel that to scrap this page would somehow break threads in the unconscious. (*A propos* of Hardy on this: he would copy out a printed proof, with possible variations; or, again, one of my arguments in our joint work.)

In 1915 I attended a 'course' for Artillery Officers. One day the afternoon was devoted to hard exercise — digging, and strenuous riding (trotting without stirrups, etc). At 4.30, without tea, we had a 'lecture', in a bleak room, with the lights in one's eyes. The 'lecture' consisted of reading from Field Service Regulations. I promptly said to myself: I don't take anything in; but it *must* be possible, by hard enough concentration, to follow at least a sentence or two. And I tried, but totally failed.

I suppose a third-class man at a University spends his entire time in this mental state.

There are people who can't sleep when deprived of a familiar night noise, e.g. street traffic. Cases are known of objections to clocks being silenced at night. There is a legend of a man who failed to sleep because the midnight gun at Durham Castle was abolished.

A very rich woman gave orders that she was not to be bothered with letters about her banking account. When £5,000 had been paid in yearly for 15 years, and not even put to current account, the Manager wrote asking if she realized the situation. 'As I can't give a plain instruction without being continually pestered, I am transferring my account elsewhere.'

Split infinitives. Once, having failed to put across that my ear stoppers were not to *hear*, but *not* to hear, I succeeded at once with 'to not hear'.

Unsplittables? To rather more than double.

My literary Ph.D. thesis: the split infinitive in Shakespeare: There is no split infinitive in Shakespeare. (He does rather markedly avoid it. Supreme case for splitting (scansion notwithstanding): To be or to not be.)

The two senses of 'un': unleash = annihilate the state of being leashed; unloose = loose into a state of un.

Lucidity is nine-tenths of style. Elisha said to the boys: if you do that again I will tell a big bear to come and eat you up. And they did. And he did. And it did. (It could do with the odd tenth.)

Ghost baths. On returning from Cornwall I rushed over to the Great Court baths. There was general dilapidation and the baths were full of mud. But having cleaned one up I had a superb hot bath. On leaving, I noticed, having missed it before, a notice in faded chalk: Out of order. On relating this in Hall: 'Oh, didn't you know, the boilers are away being cleaned.' (This is the best Ghost story *I* know.)

I dreamt (1952) that I was at a party. Catching sight of myself in a mirror, I found I was wearing a halo. My reaction was: I should hardly have thought...; but who am I to question the authorities? (This story has given great pleasure to some of my friends, e.g. Francis Clark.)

(I *had*, as it happens, done a virtuous act not long before. I was asked to dinner with a pupil: half an hour's walk there, nothing to drink, and the walk home in pouring rain.)

Virginia Woolf and 'A Room of one's own'. She had, of course, a case, but ridiculously overstated it. All very well to contrast her undergraduate's dinner at Girton with the lunch she was given in King's, but that lunch cost her host about £3 a head. And one of her statements about admission to a library was a blank lie. Literary people have no conscience at all.

About 1910 (aged 25) I passed, near Coton in Cambridgeshire, a pair of girls about 8 or 9. One had the face of an angel. Before they were out of ear-shot, angel-face said: 'I was going to say "bugger" when I saw the old man.'

Page had two stories about David Sheppard (now Bishop of Liverpool). In his first Australian tour, when he was only 19 (second year at Cambridge) and very naïve, he went in first with Hutton. Between the overs, Sheppard used to go across to Hutton and talk about God. As we know, Hutton likes to think about only one thing at a time, especially cricket and remarked 'I'm not going in first with that bugger again.' Sheppard mellowed as he grew older, but it took some time to live this down.

Apparently the second story is absolutely vouched for by teammates. Sheppard, all padded up, was found on his knees: 'Oh! Lord, Thy will be done; but a century this afternoon would be very acceptable.'

One day in a Manchester term, leaving my lodgings for the usual grim day's work, I felt an unusual sense of well-being. I presently traced this to the fact that it was not actually raining.

The tame mosquito. I suddenly became aware that a mosquito visited me each night at 6.30. It did not ever bite; but one day I killed it. And then I experienced a slight, but perceptible, pang of grief, and guilt. (Ancient Mariner?)

At Wynberg I tamed a very wild and frightened cat (it used to come and pick up food from the refuse bin): it took a long time. But then the cat was absolutely devoted. I developed a deliberately weird signature tune and used it to whistle as a signal: the cat generally came at a gallop (it would follow me on walks).

When I first heard Brahms' Serenade I was completely staggered to hear my tune, practically exact.

I have known two cases of 'two bottles of whisky a day' men. The first was in Bideford before 1914 (and the whisky much stronger). He also bathed before breakfast all the year round. The war intervened and I don't know the end, but he had a very good run for his money.

The other was Lord Howard de Walden. His doctor suggested that he really ought to cut down on the whisky; 'Why not try say two bottles of Burgundy at dinner and see if less whisky is needed?' The results were admirable.

My doctor uncle in Devon had a tired business man sent him as patient (*c.*1910). Told to eat plenty of cream, and being the kind to follow doctor's orders minutely, he called, 'I'm eating a pound with each meal; is that enough?'

Spider lines are slung across paths exactly 5 feet from the ground, because that is the height of my nose. But why do flies always go for my left ear? Some asymmetry in the use of shaving cream?

So it is only in 1947 that I learn that Aristotle said: 'When a thing has been said once it is hard to say it differently.' (After trying to write twice about Ramanujan, I have always acted up to this.)

Busoni in America, invited by a society hostess to dine in the hope that he might be prepared to play afterwards, gave the routine reply that as a professional he could do this only for a fee, which was 250 dollars. The hostess agreed that the fee would be paid, but Busoni would understand that in that case the invitation to dinner was cancelled. Busoni replied that in that case the fee would be 200 dollars. (Probably in all simplicity and with no intention to snub.)

Round about 1912 a bust was attributed to Leonardo by Bede (curator at Berlin). The Kaiser agreed. So did I, and the average amateur,

on obvious grounds of style. But K. Lucas (of Victorian date) had spe-
cialised in clever fakes; some experts attributed it to him, and on digging
out the base a Victorian petticoat (or waistcoat?) was found stuffing it.
So it was by Lucas. In 1952 it is finally attributed to Leonardo (some
trivial explanation accepted for the petticoat).

On capping stories ('whales for bait'). I took D. C. Spencer and
his wife on the walk to Land's End. Passing the curious adit hole near
Nanjizel, whose inside I described, he remarked: 'It reminds me of the
entrance to a cave in Mexico. It is rather remarkable inside. They have
got 25 miles in, but not to the end. There are stalactites 18 feet in
diameter. And a range of mountains, with lakes...'

Miscellaneous

'Always verify references.' This is so absurd in mathematics that I
used to say provocatively: 'never...'

When I began writing I innocently adopted the French habit of M.
(Monsieur) in front of any surname. I thus created a ghost mathematician M. Landau (to whom some 'non-verified' references were made).

W. S. Gilbert is rightly condemned for his brutality to plain women.
But in my experience his is nothing compared to the brutality of the
average woman *about* an unattractive man (not so much *to* him).

When Hobbes said 'Laughter was a "sudden glory"' he meant nothing poetical (though I once found that even Housman thought he did),
but that you had had an unexpected dirty score over someone.

Miracles. All that is wrong with them is that they aren't good
enough. Suppose it were announced in letters of fire across the Milky
Way that next night Sirius would shift to the middle of the Pleiades, and
that this duly happened? If it were then announced that if we did not go
to Church on Sunday we should suffer, it would be discreet to conform,
whatever philosophic reservations we might hold about the nature of the
Power concerned.

Americans have a higher standard of flattery in social intercourse
than we, and we are apt to depress them. Most people can't do the
expected thing because of feeling insincere. All this is a mistake. Have
the appropriate phrases ready in advance, and rap them out with no
attempt to *feel*; everyone will be happy.

Zygmund in Cambridge about 1928 invented the phrase 'at all' = negation of 'not at all' = 'very much no'.

Ann was fond to use 'We'd better, better'd'nt we?'

Odd reactions of the tired mind. Talking with D. S. Robertson in Hall in November about the two possible holidays from lecturing, Ash Wednesday and Ascension Day, I remarked 'It's unfortunate that the life of Christ is so uneventful in Michaelmas Term.' We had soup Pascal, and pancakes. Robertson replied: 'Surely Easter is very early this year?'

Waiting in Selfridge's for a companion to finish shopping, I found myself gazing at a thing labelled *Baby* Refrigerator. 'But why should they want to?'

I stood by, while Hardy wrote a telegram to New College in the little Post Office in Rose Crescent in Cambridge. 'Inexpressibly shocked pusillanimous surrender reactionary proposal Council stop wanton abandonment sacrosanct principle liberty, etc.' The postmistress fulfilled her duty under the regulations, and read it back with a perfectly straight face.

The Foreign Office were unable to decipher an important ms., and someone suggested a pharmacist, as being good at handwriting. 'Do you think you can read this?' — 'Oh yes, I think so.' It was returned with a bottle with the request 'That will be 2/6d.' (Via A. S. Eddington.)

Mnemonics. Starboard, right, and green are long; port, left, and red are short (and port is red).
(*Via* H. A. Hollond.) Port goes left at wine table.

Cambridge, Magdalene and Queens' are long; Oxford, Magdalen, Queen's short. (*Via* H. A. Hollond.) Summer time: Spring, jump; Fall, back.

The longest word in English, antidisestablishmentarianism, is all 'form' except for the content 'sto'.

How long does it take to cross the Atlantic at 1 knot an hour? (The answer is about 80 hours.) (The knot is a velocity.)

It is still too early to indulge in premature optimism. (C.D.N.)

To call this picture unique would be definitely an understatement. (C.D.N.)

'I think Lake Leman is more beautiful than Lake Geneva.'

'But you know Lake Leman is synonymous with Lake Geneva.'

'Maybe, but I still think Lake Leman is *more* synonymous.' (A. Lunn)

It used to be said that one of the stone balls fell off Clare Bridge whenever a Clare College man was Senior Wrangler. The proposition is true — vacuously. (It would certainly also have become true if there had been a Clare Senior Wrangler.)

I used to have a dancing-partner in Manchester 1907–10. She was Marie Stopes. We did not like each other particularly, but were good dancers, and the general standard was atrocious.

After 40 years I have now learned (1) how not to sit down on the sleeves of my gown in Hall; (2) how to light a gas fire properly; (3) that the tree in the garden under which I work is a mulberry tree; (4) that I have been tying my shoelaces with Granny knots.

Miss M. L. Cartwright, F.R.S. The only woman in my life to whom I have written twice in one day. (I would have used this in a speech at a Girton dinner in her honour, but there were no speeches.)

Harold Nicholson, in a wireless 'essay' talk between parts of a concert, objected violently to the modern sticky tape on parcels and letters. It is true that it is more blessed to give than to receive.

The small pleasures of life. Seeing a loathsome ms. transformed into beautiful print. (Comparable to watching worms on the journey towards roast duck?)

Great intellects. Does the attitude to one's schoolmaster persist? On the question who were the greatest intellects: 'Well, of course there's Newton and Archimedes... As a matter of fact the really greatest man — you wouldn't know him — he was the man who taught me algebra.' (*Via* Darwin.)

Queen Victoria's funeral. I was there. Today a royal funeral is comparable with a final deciding Australian Test Match. Queen Victoria's was comparable only with the Crucifixion, and journalists freely said so. I am glad to recall that (at 15, and though virtuous and religious minded) I was quite unimpressed — just one old lady dead.

The Danish father (via Besicovitch). Traditionally stern, he asked his wife on his deathbed why she wept. 'Because you are so kind.' — 'Not kind, only weak.'

(In 1959, I got a crawling letter from an Indian schoolmaster asking for the solution of two little problems. Such things used to go into the waste paper basket, but now I complied. The story came to my mind.)

The mongoose. In a railway carriage: *A*: what is that basket in the rack for? *B*: A mongoose; you see, I have a friend who sees snakes. *A*: But they are imaginary snakes. *B*: Oh yes, but it is an imaginary mongoose.

(This was a chestnut in 1913, and quite unknown in 1930; don't know about 1956.)

The old church. Old lady: 'We have a very old church here, 1913.' 'Oh, auntie dear, you mean 1319.' 'Not at all, my dear, 1325.' (This sharply divides the human race.)

Relativity. On the weekly tour of a set of caves, two people got separated from the party. When it became obvious they had not been missed they settled down to living for a week on their candles; they had no watches. As they were finishing the last of the 4 pieces of candle, they heard rescuers approaching. Their absence *had* been spotted, and they had been in the cave for 2 hours 40 minutes.

Riviera Express. We had booked five reserved third class seats and found them occupied (reserved twice over by mistake). We had some indecisive talk with an official in old lace. Meanwhile R. S., seeing us off, went to headquarters and threw his weight about: 'Professor Littlewood is very much annoyed.' There are only two possible lines: it is just one of those things, or complete surrender. Presently a young man in plus fours appeared: 'Professor Littlewood, it is one of the proudest moments of my life to meet you at last, I was a pupil of yours at Manchester.' We were put in a reserved First Class compartment, no-one else admitted, and lunch brought to us by special service.

Chicago. A professor, detained until after dark at the University in the winter, had to go home across the snow-covered *Midway Plaisance*. The way was by a couple of planks, and you had to wriggle past anyone going the opposite way. The professor brushed past someone, and happened at once to notice that his watch was missing. Summoning up all his courage and bravado he rushed at the man: 'Give me that watch.' To his intense relief the watch was meekly handed over. When he got home his wife said: 'You know, dear, you went out without your watch.'

Inscriptions on two gates, Alpha and Omega, on the Clovelly-Hartland road in Devon, leading nowhere. (I remember them from 1901 or so on.)

Alpha, thou'rt first, I'm sure, as Omega is in the West,

And thou'lt be first for ever more, now slumber on and rest,

This field was once a common moor, where gorse and broom
   grow free,

But now it grows green grass all o'er, as all who pass may see.

Omega, thou'rt last, I'm sure, as Alpha is in the East,

And thou'lt be first for evermore, till endless Ages cease.

When I am dead and gone

These verses will remain,

To show who wrote thereon

By working of the brain.

(Curiously fascinating; and I have remembered them effortlessly for 55
years.)

NEWTON AND THE ATTRACTION OF A SPHERE[1]

§1. In Keynes's contribution to the 'Newton Tercentenary Celebrations' there is the following passage:

'Again, there is some evidence that Newton in preparing the *Principia* was held up almost to the last moment by lack of proof that you could treat a solid sphere as though all its mass were concentrated at the centre, and only hit on the proof a year before publication. But this was a truth which he had known for certain and had always assumed for many years.

Certainly there can be no doubt that the peculiar geometrical form in which the exposition of the *Principia* is dressed up bears no resemblance at all to the mental processes by which Newton actually arrived at his conclusions.

His experiments were always, I suspect, a means, not of discovery, but always of verifying what he knew already.'

To know things by intuition happens to humbler people, and it happened, of course, to Newton in supreme degree; but I should be inclined to doubt this particular example, even in the absence of evidence. Many things are not accessible to intuition at all, the value of $\int_0^\infty e^{-x^2} dx$ for instance. The 'central' attraction of a sphere is, of course, more arguable, but in point of fact Newton says in his letter to Halley of 20th June 1686, that until 1685 (probably early spring) he suspected it to be false. (For the letter, see Rouse Ball, *An Essay on Newton's Principia*, pp. 156–159; the critical passage is from 1.7 on, p. 157. see also p. 61.) There is, I think, a sufficient and fairly plausible explanation for the proof being held up, and an analysis of the mathematical setting of the problem is of interest.

§2. I take it as established that Newton did not believe in the cen-

---

tral attraction before 1685. This being so he may well have thought the determination of the actual attraction a detail to be considered later. (It is true that in the 1666 comparison of the attraction at the moon's distance with that at the surface of the earth the departure from centrality at the surface, being at its worst, would be a serious matter. This, however, merely adds further mystery to a subject already sufficiently obscure. It is odd, incidentally, that all the accounts of the matter that I have come across before 1947 ignore this particular point.) He did, however, attack the problem in the end. Now, *with a knowledge of the answer*, the problem reduces at once to the attraction of a spherical *shell*, which in straightforward integration in cartesians happens to reduce to integration of the function $(ax + b)/(cx + d)^{3/2}$, child's play to Newton. *Without* this knowledge it is natural to attack the solid sphere; this is more formidable, and may have baffled him until he had fully developed his calculus methods. To the Newton of 1685 the problem was bound to yield in reasonable time: it is possible (though this is quite conjectural) that he tried the approach *via* a shell of radius r, to be followed by an integration with respect to r; this would of course instantly succeed. Anyhow he found a proof (and after that would always deal with the shell). But this was by no means the end of the matter. What he had to find was, as we all know, a proof, no doubt a calculus one in the first instance, which would 'translate' into geometrical language. Let the reader try.

I think we can infer with some plausibility what the calculus proof was, and I give it in modern dress. It operates, of course, on the *shell*.

*§3 Figure 24 is Newton's except that I have added the line SH and marked three angles.

Let $SH = a$, $SP = r$. The variable of integration is taken to be ϕ.[1] Consider the zone generated by rotation of HI round SP, and its (resolved) contribution $\delta F = \delta F_P$ to the total attraction F_P on P. The zone is at distance PH from P and has area $2\pi(a|\delta\theta|)HQ$; furthermore, the direction of the total attraction, $\overrightarrow{PS}$, forms an angle $\psi = \pi/2 - \theta - \varphi$ with the direction PH. Therefore the contribution is

$$(1) \qquad \delta F = \frac{1}{PH^2} \frac{PQ}{PH} 2\pi HI.HQ$$

$$= \frac{1}{PH^2} \cos\psi \, 2\pi a|\delta\theta|HQ.$$

Since $\psi = \pi/2 - \varphi - \theta$, $HQ = a\sin\theta$ and $PS/PH = \sin(\pi/2 + \varphi)/\sin\theta =$

---

[1]The order of lettering significantly corresponds to a *positive* increment $\delta\phi$.

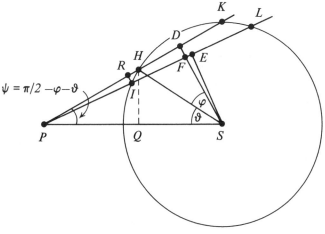

Figure 24

$\cos \varphi / \sin \theta,$

(2)
$$\frac{r^2}{2\pi a^2}\frac{dF}{d\varphi} = \left(\frac{\cos \varphi}{\sin \theta}\right)^2 \sin(\theta + \varphi) \sin \theta \left|\frac{d\theta}{d\varphi}\right|.$$

From triangle PHS we have

$$\frac{a}{r} = \frac{HS}{PS} = \frac{\sin \psi}{\sin(\pi/2 + \varphi)} = \frac{\cos(\theta + \varphi)}{\cos \varphi},$$

whence

(3)
$$-\frac{d\theta}{d\varphi} = 1 - \tan \varphi \cot(\theta + \varphi) = \frac{\sin \theta}{\sin(\theta + \varphi)\cos \varphi}.$$

From (2) and (3),

(4)
$$\frac{r^2}{2\pi a^2}\frac{dF}{d\phi} = \cos \phi.$$

The range R of ϕ is $-\frac{1}{2}\pi$ to $\frac{1}{2}\pi$, and we should arrive at the appropriate result by integration. Newton's argument, however, is in effect that R is independent of P, so that for two positions P, p we have[1] $F_P/F_p = Sp^2/SP^2$.

---

[1]This is the statement of the crucial 'Prop. 71'. The constant is actually never determined.

§4. Now for the geometrical proof (which must have left its readers in helpless wonder). There is a counterpart figure in small letters (the two spheres being equal). The use of ϕ-integration translates into the 'peculiar' idea 'let phk cut off an arc hk equal to HK, and similarly for pil'. The details that follow are in essentials elegant and very carefully arranged, but the archaic language makes for heavy going, and I modernise. The *difficulties* arise from (3), naturally troublesome to translate.

Because of (1), the contributions $\delta F_{P,p}$ of the two respective incremental zones of HI, hi satisfy

(5)
$$\frac{\delta F_p}{\delta F_P} = \frac{PH^2}{ph^2}\left(\frac{pq}{ph}\Big/\frac{PQ}{PH}\right)\frac{hi}{HI}\frac{hq}{HQ}$$
$$= \frac{PH^2}{ph^2}\left(\frac{pd}{ps}\Big/\frac{PD}{PS}\right)\frac{hi}{HI}\frac{hq}{HQ}.$$

Since $hk = HK$ and $il = IL$, we have $sd = SD$, $se = SE$ and $df = sd - se = SD - SE = DF$. Therefore, as $\angle RHI = \angle rhi$,

(6)
$$\frac{pd.PH}{ph.PD} = \frac{pd}{pr}\Big/\frac{PD}{PR} = \frac{df}{ri}\Big/\frac{DF}{RI} = \frac{RI}{ri} = \frac{IH}{ih}$$

and

$$\frac{PH}{HQ}\Big/\frac{ph}{hq} = \frac{PS}{SD}\Big/\frac{ps}{sd} = \frac{PS}{ps}.$$

Multiplying (6) and (7) we have

$$\frac{PH^2.pd.hq}{ph^2.PD.HQ} = \frac{IH.PS}{ih.ps},$$

which combines with (5) to give

$$\frac{\delta F_p}{\delta F_P} = \frac{PS^2}{ps^2}.$$

§5. There is a proof satisfying Newton's canons, which one feels he might have arrived at if he had found no other way. It arises easily enough out of the modern approach and is as follows: it operates on the solid sphere. (The original notation P, S is being continued.)

Consider $\sum$, the concentric sphere through P, the component attraction $N_Q(P)$ at P normal to $\sum$ of a unit particle at Q, and the average $\overline{N}_Q$ of $N_Q(P)$ over all P of $\sum$. The contribution to the total $4\pi a^2 \overline{N}_Q$, i.e. average multiplied by area, of an element of area $\delta \sum$ is, by easy geometry, the solid angle subtended by $\delta \sum$ at Q. Hence $4\pi a^2 \overline{N}_Q = 4\pi$, and $\overline{N}_Q$ is independent of Q and so equal to $\overline{N}_S$. Now $\sum \overline{N}_Q \delta V_Q$, taken over elements of volume of the solid sphere, is the average normal attraction of the solid sphere taken over points of $\sum$, and its equivalent $\overline{N}_S \sum \delta V_Q$ is the corresponding thing for the mass concentrated at S. In each case the thing averaged is constant and equal to the *total* force, and the equivalence gives what we want.

§6. Return to the question of 'intuition'. The obvious argument 'against' is: 'What has the inverse square got that the inverse cube hasn't?' There is an answer to this: the inverse square is the 'natural' law of diminution with distance, e.g. of light or sound, and others than Newton thought of it in connexion with the planetary system.[1]

Alternatively, adopt a corpuscular theory of light, in which corpuscles from different origins do not collide. If they are, say, inelastic, the corpuscles from a point create a *repulsion* according to the inverse square, and *the total pressure on a sphere with the point as centre is independent of the radius.* This independence, combined with the symmetry, might produce the feeling that the situation in the original attraction problem is best compatible with 'centrality': an intuition. But by this time we are within striking distance of the proof given above. The total outward (i.e. normal) pressure on a $\sum$ due to an origin of corpuscles at Q is a not unnatural idea, and the key is to prove that it is independent of the position of Q.*

---

[1] It may be observed that with a law r^{-0} the attraction at the surface of the solid sphere is less than the central value, while with the law r^{-4} it is greater. These facts are obvious, in the second case because the attraction concerned is infinite.

THE DISCOVERY OF NEPTUNE

Neptune was discovered in 1846 as a result of mathematical calculation, done independently and practically simultaneously by Adams and le Verrier. The full story abounds in unexpected twists, and is complicated by personal matters, some of them rather painful. There is a fascinating account in Professor W. M. Smart's *John Couch Adams and the Discovery of Neptune*, published by the Royal Astronomical Society, 1947. I am concerned only with limited parts of the field.

To refresh the reader's memory of what has been said from time to time about the discovery I will begin with some representative quotations. In *The Story of the Heavens* (1886) Sir Robert Ball has passages: 'the name of le Verrier rose to a pinnacle hardly surpassed in any age or country' ... 'profound meditations for many months' ... 'long and arduous labour guided by consummate mathematical artifice.' The author is not above a bit of popular appeal in this book — 'if the ellipse has not the perfect simplicity of the circle, it has at least the charm of variety ... an outline of perfect grace, and an association with ennobling conceptions' — but on Neptune he is speaking as a professional. An excellent modern book on the history of astronomy has, so late as 1938: 'probably the most daring mathematical enterprise of the century ... this amazing task, like which nothing had ever been attempted before.'

The immediate reaction was natural enough. Celestial Mechanics in general, and the theory of perturbations in particular, had developed into a very elaborate and highbrow subject; the problem of explaining the misbehaviour of Uranus by a new planet was one of 'inverse' theory, and the common feeling was that the problem was difficult up to or beyond the point of impossibility. One might speculate at some length on reasons for this opinion (one, perhaps, was confusion between different meanings of the technical term 'insoluble[1]). When Adams and le Verrier

---

[1]Its attachment to the '3-body problem' misleads people today.

proved the opinion wrong (and after all *any* mathematical proof is a debunking of sorts) there was still something to be said for the principle that difficulties are what they seem before the event, not after. Certainly no one would grudge them their resounding fame. (Nor grudge, at a lower level, the luck of a discovery which makes a more sensational impact than its actual difficulty strictly merits; in point of fact this luck never does happen to the second-rate.) If the discovery has had a very long run one must remember that there is a time-lag; people cannot be always reconsidering opinions, and having said something once even the most intelligent tend to go on repeating it. The phrase was still in vogue that 'only 3 people understand Relativity' at a time when Eddington was complaining that the trouble about Relativity as an examination subject in 'Part III' was that it was such a soft option.

In what I am going to say I am far from imputing stupidity to people certainly less stupid than myself. My little *jeux d'esprit* are not going to hurt anyone, and I refuse to be deterred by the fear of being thought disrespectful to great men. I have not been alone in a lurking suspicion that a much simpler approach might succeed. On the one hand, aim at the minimum needed to make observational discovery practicable, specifically aim at the time t_0 of conjunction.[1] On the other, forget the high-brow and laborious perturbation theory, and try 'school mathematics'. (I admit to the human weakness of being spurred on by the mild piquancy success would have.) To begin with I found things oddly elusive (and incidentally committed some gross stupidities). In the end an absurdly simple line emerged: I can imagine its being called a dirty trick, nor would I deny that there is some truth in the accusation. The only way to make my case is to carry out the actual 'prediction' of t_0 from the observational data, with all the cards on the table (so that anyone can check against unconscious — or conscious — faking). I will also write so as to take as many amateurs as possible with me on the little adventure.

A planetary 'orbit' is an ellipse with the Sun S at a focus, and the radius vector SP sweeps out area at a constant rate (Kepler's second law). An orbit, given its plane, is defined by 4 elements, a, e, α, ϵ. The first 3 define the geometrical ellipse: a is the semi-major axis; e the eccentricity; α the longitude of perihelion, i.e. with the obvious polar coordinates r, θ, θ is the 'longitude' and $\theta = \alpha$ when P is nearest S (at an end of the major axis). When we know a we know the 'mean angular velocity' n and the associated period $p = 2\pi/n$; n is in fact proportional

---

[1]The time at which NUS is a straight line (I shall use the abbreviations S, U, N).

to $a^{-3/2}$ (Kepler's third law)[1]; further the constant rate of area sweeping is $\frac{1}{2}abn$,[2], and *twice* this rate is identical with the 'angular momentum'[3] (a.m. for short); this has the differential calculus formula $r^2\dot{\theta}$, and it also is of course constant (Kepler's second law). The 4th element, the 'epoch' ϵ, is needed to identify the origin of t; the exact definition is that $\theta = \alpha$ (perihelion) occurs at the t for which $nt + \epsilon = \alpha$.

Uranus' orbit has a period of 84 years, and an eccentricity e of about $\frac{1}{20}$. The effects of bodies other than the Sun and Neptune can be allowed for, after which we may suppose that Uranus, the Sun, and the eventual Neptune are the only bodies in the system; we may also suppose (all this is common form) that the movements are all in one plane. The values of θ (for Uranus) at the various times t (we sometimes write $\theta(t)$ to emphasize that θ is 'at time t') may be regarded as the observational raw material though of course the actual raw observations are made from the Earth. The r's for the various t are indirect and are much less well determined.

The position in 1845 was that no exact elliptic orbit would fit the observed θ over the whole stretch 1780 to 1840.[4] The discrepancies are very small, mostly within a minute of arc (with a sudden swoop of about $90''$, see Table 1). The ratio m of Neptune's mass to that of the Sun (taken as 1) is actually about 1/19000 (the orders of magnitude fit since m radians is about $11''$).

In the absence of Neptune the a.m. A is constant (as stated above — *alias* of Kepler's second law); *the actual Neptune accelerates A at times earlier than t_0 and decelerates it at later times.* The graph of A against t therefore rises to a maximum at $t = t_0$, and my first idea was that this would identify t_0. So it would if all observations were without error (and the method would have the theoretical advantage of being unaffected by the eccentricities). But the value of A at time t depends on the r at time t, and the determinations of the A's are consequently too uncertain. Though the method fails it rises from the ashes in another form. For this a few more preliminaries are needed.

The numerical data Adams and le Verrier had to work on were not the observed θ's themselves, but the differences between the observed $\theta(t)$

---

[1]It does *not* depend on e.

[2]The total area of the ellipse is πab, and it is swept out in time p.

[3]Strictly speaking the a.m. should have the planet's mass as a factor; but U's mass is irrelevant and I omit it throughout.

[4]Observations after 1850 were not immediately available, and anyhow were not used. Uranus was discovered in 1781. Lest the reader should be worried by small inconsistencies in the dates I mention that 1780 is 'used', the extrapolation being a safe one.

TABLE 1

Year	Observed δ	Smooth Curve	Year	Observed δ	Smooth Curve
1780	3·5	3·5	1813	22·0	22·8
1783	8·5	8·5	1816	22·9	22·5
1786	12·4	12·5	1819	20·7	22·0
1789	19·0	15·8	1822	21·0	21·0
1792	18·7	18·3	1825	18·2	18·1
1795	21·4	20·3	1828	10·8	10·3
1798	21·0	21·6	1831	− 4·0	− 4·0
1801	22·2	22·4	1834	−20·8	−20·8
1804	24·2	22·8	1837	−42·7	−42·5
1807	22·1	23·0	1840	−66·6	−66·6
1810	23·2	23·0	1843(e)	—	−94·0

and the $\theta_B(t)$ of an elliptic orbit calculated by Bouvard; the 'discrepancy' $\delta(t)$ (δ for short) is $\delta(t) = \theta(t) - \theta_B(t)$. ($\theta_B(t)$ depends on the 'elements' of E_B, and these are subject to 'errors'. These errors are among the unknowns that the perturbation theory has to determine: our method does not mind what they are, as we shall see.) Table 1 gives the raw δ's (given in Adams's paper[1]) together with the values got by running a smooth curve. The treatment of the start of the sudden swoop down after the long flat stretch is a bit uncertain: I drew my curve and stuck to it (but faking would make no ultimate difference). The differences show up the order of the observational errors (which naturally improve with the years — something seems to have gone badly wrong for 1789); these are absolute, not relative (thus the probable absolute error in $\delta_1 - \delta_2$ is the same whether $\delta_1 - \delta_2$ is 0·5″ or 90″). It is worthwhile to work to 0·1″ and to the number of decimal places used in what follows, even though the last place is doubtful.

The value for 1843 is an extrapolation; results derived from it are labelled (e)'.

An 'effect' due to Neptune is of 'order m, in mathematical notation $O(m)$; if, for a particular quantity X, ΔX denotes (calculated X)−(observed X), then any ΔX is $O(m)$. The square of this (2nd order of infinitesimals) is extremely minute and everyone neglects it instinctively (if a watch loses 10 seconds a day you don't try to correct for the further

---

[1]Collected Works, I, p. 11. These (and *not* the modifications he introduces, which are what appear in Smart) are what is relevant for us. The δ's are actually 3-year averages.

loss over the lost 10 seconds — the cases are comparable). Next, an effect of Neptune is what it would be if Uranus, and also Neptune, moved in circles, *plus* a 'correction' for the actual eccentricities of the orbits. Uranus' eccentricity e $\left(\frac{1}{20}\right)$ is unusually large and it would be reasonable to expect Neptune's to be no larger (it is actually less than $\frac{1}{100}$). The e's distort the 'circular' value of the effect by 5 per cent (or say a maximum of 10 per cent.); the 'distortion' of the effect is $O(em)$, the effect itself being $O(m)$. I propose to ignore things of order $O(em)$[1]: this is the first step in my argument. In particular, when we have something which is either some Δ, or m itself, multiplied by a *factor*, we can substitute first approximations (i.e. with $e = 0$), or make convenient changes that are $O(e)$, in the *factor*.

Suppose now that E_1, E_2 are two (exact) elliptic orbits, yielding $\theta(t)$'s that differ by amounts of the kind we are concerned with, differing, that is, by $O(m)$.[2] It is now the case that *the differences satisfy the equation*

$$(1) \qquad \theta_1 - \theta_2 = m(a + bt + c\cos nt + d\sin nt) + O(em),$$

where a, b, c, d are constants depending on the two sets of elements of E_1, E_2, and (following our agreement about *factors* of m) n is any common approximation to the mean angular velocity. I will postpone the school mathematics proof of this.

Next, (i) let E^* be the 'instantaneous orbit at time t_0', that is to say the orbit that Uranus would describe if Neptune were annihilated at time t_0 : note that E^* shares with t_0 the property of being 'unknown'. (ii) Let ϑ be the perturbation of the θ of Uranus produced by Neptune *since time t_0*[3] Then if, at any time t, θ is (as usual) Uranus' longitude, θ_B is the longitude in the orbit E_B and θ^* the longitude in the orbit E^*, we have $\vartheta = \theta - \theta^*$, and so

$$(2) \qquad \delta(t) = \theta - \theta_B = (\theta^* - \theta_B) + \vartheta.$$

Now both the last terms have a factor m, and we may omit any stray $O(em)$'s. In particular, we may in calculating ϑ drop *any* e terms. But

---

[1] I should stress that there is no question of ignoring even high powers of e *unaccompanied by a factor m* (e^4 radians is about $1''$). The distortion in the value found for t_0 is, however, a sort of exception to this. But the effect of e's in distorting t_0 is unlikely to be worse than the separation they create between time of conjunction and time of closest approach. An easy calculation shows that this last time difference is at worst 0·8 years.

[2] The orbits may have 'Suns' of masses differing by $O(m)$.

[3] We allow, of course, *negative* values of $t - t_0$ both in E^* and in ϑ.

this means that *we can calculate ϑ as if both Uranus' and Neptune's orbits were circles*. When, however, the orbits are circles, ϑ has equal *and opposite values at t's on equal and opposite sides of t_0*; in other words, *if we write $t = t_0 + \tau$, then*

$$(3) \qquad\qquad \vartheta(t) = \Omega(\tau),$$

where $\Omega(\tau)$ is an odd[1] function of τ; i.e. $\Omega(-\tau) = -\Omega(\tau)$.

This, used in combination with (1) and (2), is the essential (and very simple) point of the argument. The difference $\theta^* - \theta_B$ is a special case of $\theta_1 - \theta_2$ in (1). Write $t = t_0 + \tau$ in (1) and combine this with (2) and (3); this gives, ignoring $O(em)$'s,

$$\delta(t_0 + \tau) = m\left\{a + bt_0 + b\tau + c\cos(nt_0 + n\tau) + d\sin(nt_0 + n\tau)\right\} + \Omega(\tau).$$

Expanding the cos and sin of sums and rearranging we have (with new constants, whose values vary with t_0 but do not concern us)

$$\delta(t_0 + \tau) = A - B(1 - \cos n\tau) + \left\{C\tau + D\sin n\tau + \Omega(\tau)\right\}.$$

The curly bracket is an odd function of τ. Hence if we combine equal and opposite τ and construct $\delta^*(\tau)$ and $\rho(\tau)$ to satisfy

$$\delta^*(\tau) = -\tfrac{1}{2}\left\{\delta(t_0 + \tau) + \delta(t_0 - \tau) - 2\delta(t_0)\right\},$$
$$\rho(\tau) = \delta^*(\tau)/(1 - \cos n\tau),$$

we have $\delta^*(\tau) = B(1 - \cos n\tau)$, and so $\rho(\tau) = B$ for all τ. *If, then, we are using the right t_0, the ratio $\rho(\tau)$ must come out constant*: this is our method for identifying t_0. The actual value of t_0 to the nearest year is 1822.

Table 2, in which the unit of time is 1 year (and the n of $\cos n\tau$ is $2\pi/84$), shows the results of trying various t_0 (the century is omitted from the dates). The last place of decimals for the $\rho(\tau)$ is not reliable, but of course gets better as the size of the entry $\delta^*(\tau)$ increases: I give the numbers as they came, and they speak for themselves. $\tau = 6$ is included, though the proportionate error in δ^* is then considerable.[2] For $t_0 = 13$ ρ goes on to 34·8 at $\tau = 27$; for $t_0 = 16$ it goes to 38·2 at $\tau = 24$. Once

---

[1] 'Ω' is a deputy for 'O' (initial of 'odd'), which is otherwise engaged.

The italicized statement in the text is true 'by symmetry': alternatively, *reverse the motions from time t_0*. (The argument covers also the perturbation of the *Sun*, which is not so completely negligible as might be supposed.)

[2] And the values for $\tau = 6$ at $t_0 = 22$, 22·4 are more uncertain than usual because of a crisis in the smooth curve.

TABLE 2

τ	$t_0 = 13$		$t_0 = 16$		$t_0 = 19$		$t_0 = 22$	
	$2\delta^*(\tau)$	$\rho(\tau)$	$2\delta^*(\tau)$	$\rho(\tau)$	$2\delta^*(\tau)$	$\rho(\tau)$	$2\delta^*(\tau)$	$\rho(\tau)$
6	0·6	3·0	1·0	5·1	3·6	18·3	9·2	47·0
9	1·8	4·1	3·9	9·0	10·7	24·6	23·2	53·3
12	5·3	7·2	11·9	15·8	25·0	33·2	39·8	53·0
15	13·7	12·1	26·6	23·5	42·0	37·1	61·5	54·4
18	29·3	18·8	44·2	28·4	64·1	41·2	85·8	55·2
21	48·3	24·1	67·2	36·6	89·0	45·5	113(e)	56·5(e)

τ	$t_0 = 22·4$		$t_0 = 25$		$t_0 = 28$	
	$2\delta^*(\tau)$	$\rho(\tau)$	$2\delta^*(\tau)$	$\rho(\tau)$	$2\delta^*(\tau)$	$\rho(\tau)$
6	10·6	53·8	18·2	92·5	20·4	103
9	25·0	57·6	34·5	79·3	41·1	94
12	42·2	56·1	55·9	74·4	64·7	86
15	64·0	56·5	79·8	70·5	91(e)	80·6(e)
18	88·6	57·0	106(e)	68·3(e)	—	—
21	116(e)	58·0(e)	—	—	—	—

the data — the smooth curve values — were assembled the calculations took a mere hour or so with a slide-rule. The date 1822·4 seems about the 'best' t_0.

We need fairly large τ for $\delta^*(\tau)$ to have enough significant figures, and further to provide a range showing up whether $\rho(\tau)$ is constant or not. And we need room to manoeuvre round the final t_0. So the method depends on the 'luck' that 1822 falls comfortably inside the period of observation 1780–1840. But *some* luck was needed in any case.

It is an important point that the method is quite indifferent to how well E_B does its originally intended job, and *we do not need to know (and I don't know) its elements*; it is enough to know the 'discrepancy' with *some*, 'unknown', orbit (not *too* bad of course). On the other hand the method ostentatiously says nothing at all about the mass or distance of Neptune. I will add something on this. With e-terms ignored $\vartheta(\tau)/m$ can be calculated *exactly* for any given value of $\lambda = a/a_1$ (ratio of the a's of Uranus and Neptune).[1] The idea would be to try different λ's,

---

[1] From two second order differential equations. The formula involves 'quadra-

each λ to give a best fitting M, and to take the best fitting *pair* λ, m. This fails, because the greater part of ϑ is of the form $b'(n\tau - \sin n\tau)$, and b' is smothered by the a, b, c, d of $\theta^* - \theta_B$ which depend on the unknown elements of E_B (ϑ is smothered by the 'unknown' $\theta^* - \theta_B$). If we *knew* these elements (or equivalently the raw θ) we might be able to go on. They could be recovered from the Paris Observatory archives; but this article is a last moment addition to the book, I do not feel that I am on full professional duty, and in any case we should be losing the light-hearted note of our adventure.

The time t_0 once known, it would be necessary to guess a value for Neptune's distance a_1; Neptune's period is then $84(a_1/a)^{3/2}$ years, and we could 'predict' Neptune's place in 1846. The obvious first guess in 1846 was $a_1/a = 2$, following Bode's empirical law, to which Neptune is maliciously the first exception, the true value being 1·58. Adams and le Verrier started with 2 (Adams coming down to 1·942 for a second round). Since from our standpoint[1] too large an a_1 has disproportionately bad results as against one too small, it would be reasonable to try 1·8. This would give a prediction (for 1846) about 10° out, but the sweep needed would be wholly practicable.

Le Verrier was less than 1° out (Adams between 2° and 3°); 'they pointed the telescope and saw the planet'. This very close, and double, prediction is a curiosity. All the observations from 1780 to 1840 were used, and on an equal footing, and the theory purported to say where Neptune was over this whole stretch. With a wrong a_1 they could be right at 1840 only by being wrong at 1780. With Adams's $a_1 = 1\cdot 94a$ Neptune's period (which depends on a_1 only) would be 227 years; he would have been wrong by 30° for 1780 if the orbit were circular, and so the angular velocity uniform. But faced with a wrong a_1 the method responded gallantly by putting up a large eccentricity $(\frac{1}{8})$, and assigning perihelion to the place of conjunction. The combination makes the effective distance from S over the critical stretch more like $1\cdot7a_1$, and the resulting error at 1780 (the worst one) was only 18°. (A mass 2·8 times too large was a more obvious adjustment.)

In much more recent times small discrepancies for Neptune and Uranus (Uranus' being in fact the more manageable ones) were analysed for a trans-Neptunian planet, and the planet Pluto was found in 1930 near the predicted place. This was a complete fluke: Pluto has a mass probably no more than $\frac{1}{10}$ of the Earth's; any effects it could have on

---

tures', but in numerical calculation integration is quicker than multiplication. It would be comparatively easy to make a double entry table for $\vartheta(\tau, \lambda)/m$.

[1]Perturbation theory calculations have necessarily to *begin* by guessing an a_1; our guess need only be at the end.

Neptune and Uranus would be hopelessly swamped by the observational errors.

It remains for me to give the (school mathematics) proof of (1) above. Call $e_1 - e_2$ Δe, and so on. I said above that *all* Δ's were $O(m)$: this is not quite true, though my deception has been in the reader's best interests, and will not have led him astray. It is true, and common sense, for Δa, Δe, Δn, and $\Delta \epsilon$. But the 'effect' of a given $\Delta \alpha$ vanishes when $e = 0$, and is proportional to e. *It is $e\Delta\alpha$, not $\Delta\alpha$ that is comparable with the other Δ's and so is $O(m)$.*[1]

We start from two well-known formulae. The first is geometrical; the polar equation of the ellipse of the orbit is

$$(4) \qquad r = a(1 - e^2)\left((1 + e\cos(\theta - \alpha))\right)^{-1}.$$

The second is dynamical; the equation of angular momentum (Kepler's second law) is

$$(5) \qquad r^2 \frac{d\theta}{dt} = na^2(1 - e^2)^{1/2}.$$

So, using dots for time differentiations,

$$(6) \quad \dot\theta = n(1 - e^2)^{-3/2}\left[1 - 2e\cos(\theta - \alpha) + 3e^2\cos^2(\theta - \alpha) + \ldots\right].$$

The first approximation (with $e = O$) is $\theta = nt + \epsilon$. We take suffixes 1 and 2 in (6) and operate with Δ, remembering that we may take first approximations in any *factor* of an m.

In estimating $\Delta\dot\theta$ we may, with error $O(em)$, ignore the factor $(1 - e^2)^{-3/2}$ in (6), since it is itself $1 + O(e^2)$, and its Δ is $O(e\Delta e) = O(em)$. We have, therefore, with error $O(em)$,

$$\Delta\dot\theta = \Delta\{n[\ ]\} = [\ ]\Delta n + n\Delta[\ ].$$

The 1st term is $\Delta n + O(em)$. The 2nd is

$$n\left[\Delta e\{-2\cos(\theta - \alpha)O(e)\} + \Delta(\theta - \alpha)\{2e\sin(\theta - \alpha) + O(e^2)\}\right],$$

---

[1]This twist makes the 'obvious' approach of using the well-known expansion

$$\theta = nt + \epsilon + 2e\sin(nt + \epsilon - alpha) + \tfrac{5}{4}e^2\sin 2(nt + \epsilon - \alpha) + \ldots$$

slightly tricky; we should have to keep the term in e^2. The line taken in the text side-steps this.

and we may drop the θ in $\Delta(\theta - \alpha)$ on account of the factor $O(e)$. Summing up, we obtain

$$\Delta\dot\theta = m\left(A + B\cos(\theta - \alpha) + C\sin(\theta - \alpha)\right) + O(em),$$

where $mA = \Delta n$, $mB = -2n\Delta e$, $mC = -2n(e\Delta\alpha)$. Substituting the first approximation $\theta = nt + \epsilon$ in the right hand side, we have

$$\Delta\dot\theta = m\left(A + B\cos(nt + \epsilon - \alpha) + C\sin(nt + \epsilon - \alpha)\right) + O(em),$$

and integration then gives

$$\Delta\theta = \Delta\epsilon + m\left(At + (B/n)\sin(nt + \epsilon - \alpha)\right.$$
$$\left. - (C/n)\cos(nt + \epsilon - \alpha)\right) + O(em),$$

which, after expanding the sin and cos and rearranging, is of the desired form (1).[1]

---

[1] We have treated Δn and Δa as independent (the latter happens not to occur in the final formula for $\Delta\theta$): this amounts to allowing different masses to the two 'Suns'. The point is relevant to certain subtleties, into which I will not enter.

THE ADAMS–AIRY AFFAIR

Synopsis.[1] Adams called at the Royal Observatory on Oct. 21, 1845, failed to see Airy, and left a note with a short statement of his predictions about Neptune. Airy's letter of reply (Nov. 5) contained a question 'whether the errors of the radius vector would be explained by the same theory that explained the errors of longitude'. Adams did not reply. The observational search did not begin till July 29, 1846, when Challis (in Cambridge) embarked on a comprehensive programme of sweeping (unfortunately much too comprehensive) that continued to the end of September. Le Verrier sent his predictions to the Berlin Observatory, where Galle (aided by a recently published star map) found Neptune on September 23, the day le Verrier's letter arrived.

What I have to say centres round these '*r*-corrections'. Adams started with a firm belief that the cause of Uranus' misbehaviour was an unknown planet, and a sure insight into how to carry through the mathematics; he was fully concentrated on the job — 'on duty'. Airy, not on duty, thought, not at all unreasonably in 1845, that there might be all sorts of other possibilities in an obscure field, and was sceptical about a new planet. (And probably with a background, based on past analogies, that perturbation methods would call for observational material over several revolutions — many hundreds of years.) He did, however, raise the question about *r*-corrections, and this unfortunately became a pet idea. (Airy, at the R.A.S. meeting, Nov. 13, 1846 'explained' 'I therefore considered that the trial, whether the error of the radius vector would be explained by the same theory which explained the error of longitude, would be truly an *experimentum crucis*. And I waited with much anxiety for Mr. Adams's answer to my query. Had it been in the affirmative, I should have at once exerted all the influence etc'.) Adams (who said no word from first to last that failed in courtesy or generosity) did in fact

---

[1]The full story occupies the greater part of Smart: *John Couch Adams and the Discovery of Neptune* pp. 19–43.

not answer. His private reason[1] was that he thought the question trivial. What he says to Airy (in reply, Nov. 18, to the quotation above), tactful apologies omitted, is as follows. 'For several years past the observed place of Uranus has been falling more and more behind its tabular place. In other words the real angular motion of Uranus is considerably *slower* than that given by the tables. This appeared to me to show clearly that the tabular radius vector would be considerably increased by any theory which represents the motion in longitude, for the variation in the second member of the equation

$$r^2 \frac{d\theta}{dt} = \sqrt{(\mu a(1 - e^2))}$$

is very small.[2] Accordingly I found that if I simply corrected the elliptic elements so as to satisfy the modern observations as nearly as possible, without taking into account any additional perturbations, the corresponding increase in the radius vector would not be very different from that given by my actual theory'. (The rest is irrelevant to the matters at issue.)

I find this (like so much written up to say 80 years ago[3]) very far from a model of lucidity; but the essential point, that the a.m. varies very little, could not be clearer. and 'constancy' of the a.m. would establish a simple linkage between the θ-errors and the r-errors. Challis (after the discovery) agrees: 'It is quite impossible that (Uranus') longitude could be corrected during a period of at least 130 years independently of the correction of the radius vector...' Adams is finally confirmed in his view by the actual numerical calculations; the variation of the a.m. provides only a small contribution to the θ-effects.

The first thing to note is that Airy and Adams are partly at cross-purposes. Airy's background is: I doubt the explanation by a planet, but would take it up if it explained the r's as well as the θ's. Adams's is: a planet is the explanation, and if it is determined to fit the θ's it can't help (on account of the 'linkage') fitting the r's too. Adams was clearly at fault in not seeing and allowing for this; he did later admit that not giving a reasoned answer at once amounted to a lapse on his part. It is probably relevant that besides being the fine mathematician he was Adams happened also to be very much the 'Senior Wrangler type' (an extreme form, on its own ground, of the bright young man); knowledge

---

[1] Given in a conversation with Glaisher in 1883.

[2] ($r^2 \frac{d\theta}{dt}$ is the angular momentum (a.m. for short). J.E.L.)

[3] A man had to be of 'Fellowship standard' to read a paper with understanding; today a marginal Ph.D. candidate can *read* anything.

and ideas in pigeon holes available at any moment; effortlessly on duty. Older people, off duty whenever possible, and with pigeon holes mislaid, can seem slow and stupid.[1] Always answer the 'trivial' questions of your elders (and it is just possible even for a bright young man to be overlooking something).

I have been keeping up my sleeve the fact that on the crucial theoretical point about the a.m. Adams was dead wrong. The variation in the a.m. is proportionately of the same order as the θ-error, This is obvious from the point of view of school mathematics (take 'moments' about the Sun: Neptune pulls hard at the a.m.).[2] Adams's point of view was consistently perturbation theory, but even so, and granted that even a Senior Wrangler type can make a slip, it is an odd slip to make.[3] The numerical confirmation is a last touch of comedy. In 'small contribution' 'small' means small in the sense that 15 per cent is small, not 'very' small (Adams's word). But in any case this 'smallness' is an accident of the numerical constants; e.g. it does not happen for a 'distant' Neptune.

Postscript on celestial mechanics. I will wind this up by debunking a recent piece of work of my own, which has as a consequence that *a gravitating system of bodies* (a generalization of the Solar System) *can never make a capture, even of a speck* (or, reversely, suffer a loss).[4] This is 'sensational' (I have not met a mathematician who does not raise his eyebrows), and contrary to some general beliefs. I should add: (i) it is not that the speck promptly goes out again; it may be retained for any number of billion years. And there are *limiting* cases in which the capture is permanent; but these are to be rejected as being infinitely rare, just as we reject a permanent state of unstable equilibrium (a pin on its point). These infinitely rare happenings, however, show that the

---

[1]No one who does not meet eminent people off duty would credit what they are capable of saying. I recall two conversations at the Trinity High Table. One of the most eminent of biologists was asked whether two sons, one from each of two marriages, of identical twin brothers to identical twin sisters, would be identical; he replied 'yes', and was corrected by a philosopher whose pigeon holes are always in unusually good working order. In the other Rutherford, Fowler, at least one other physicist, and myself got into a hopeless muddle between the alleged 'penny and feather in a vacuum' experiment and the fact that viscosity is independent of density. Was the experiment a bluff? Rutherford said apologetically that he *thought* he had seen it done as a boy. The muddle continued until after dinner we were put out of our misery by an engineer.

[2]The full analysis of the circular case of course confirms this.

[3]$1 - e^2$ has 'small variation', but what makes him think that a has?

[4]The bodies are idealized to be point-masses, to avoid bumps arising from finite size, and subject to Newton's inverse square law of gravitation (this is probably not essential). A number of inquiries have failed to disclose any previous statement of the result.

theorem cannot be trivial. (ii) The proof in no way shows that it is the speck that goes out, it might be Jupiter.

The theorem is a consequence of the following one: *suppose a system has been contained within a fixed sphere the Sun for all negative time, then (unless it is one of an infinitely rare set) it will be so contained for all positive time. Similarly with positive and negative times reversed.* (To see that the former theorem *is* a consequence, observe that any *genuine* capture, or escape, must involve a difference between past and future time that is ruled out.)

Take the new theorem in its (slightly more convenient) reverse form: the ideas for proving it are as follows. A 'system' is associated with a representative point (r.p. for short), P say, of a 'phase space', embodying the 'initial conditions 'at a fixed time t_0 say $t_0 = 0$.[1] Now take the 'set' V (in the $6n$-space) of *all* points P representing systems that (in the astronomical space) stay in the sphere the Sun for all positive time. The 'system' P has a new r.p. P' (coordinates $x, \ldots, \dot{x}, \ldots$), at time (say) $t = 1$, and to the set V of P's there corresponds a set V' of P''s. A well-known theorem, which we will take for granted, says that, because the differential equations for the system are 'conservative', the ($6n$-dimensional) volumes of V and V' are equal.[2] Next, consider V' as in the *same* $6n$-space as V. For a r.p. P to belong to V it has to satisfy a certain entrance examination, namely that all its bodies should stay in the Sun in the *future* of the system. Now any P' of V' is derived from some P of V, P''s 'future' starts time 1 later than P's, and its bodies stay in the Sun; P' *satisfies the entrance examination.* So: *the set V' is contained in V.* But the volumes of V' and V are equal. The two things together clearly require the sets V' and V to be *identical* (as *wholes*; the *collection* of P''s is the same as the collection of P's). So *a point Q of V is some point or other of V', R say.*

Now start with a point Q of V, and take the corresponding r.p. at time $t = -1$ (time 1 into the past). Q is an R of V', and is therefore 'derived' (i.e. as P' is 'derived' from P) from some T of V. This T *is the r.p. of P at $t = -1$. So: the r.p. at $t = -1$ of a P of V is itself in V.* Now this repeats *indefinitely* into the past: if P is in V at $t = 0$ the corresponding r.p. is in V at times (of the form $t = -m$) going back into the infinite past. This means that its bodies stay in the Sun in the *whole* of the infinite past, and we have arrived.

---

[1] If there are $n + 1$ bodies the phase space has $6n$ dimensions, the 'coordinates' being the (astronomical) space coordinates x_0, y_0, z_0 and the corresponding velocities $\dot{x}_0, \dot{y}_0, \dot{z}_0$ of n of the bodies at time t_0.

[2] *Any* set V has the same volume at all t. For the professional there is a 1-line proof (if you can call Jacobian determinants a line).

This argument is an astonishing example of the power of general reasoning. If the ideas in it were my own I should indeed have done something; but they are at least 60 years old. What happened to *me* was this. I had been lecturing for some time on differential equations of a kind for which the volume corresponding to V *decreases* with increasing t. The 'constant volume theorem' (which in my innocence I had learned only recently) came to mind, and I switched over to the Celestial Mechanics equations by way of a change. During a stroll after the lecture the argument I have given flashed through my mind (literally in a matter of seconds). My first feeling was that I could not publish a thing which had so little originality. But finding myself hopelessly behindhand with a promised contribution to a *Festschrift* I began considering details. It is the way people think who ever think of anything new at all, but taken strictly the argument contains some lies (3 to be exact). To straighten these out is a job any competent analyst could do, but it puts up a colourable appearance of backbone. So the *trouvaille* was written up[1]: brilliant ideas, not mine, *plus* a routine job.

---

[1] On the problem of n bodies, *Communications du séminaire mathématique de l'Université de Lund, tome supplémentaire (1952), dédié à Marcel Riesz*, 143–151; Corrections, Kunl. Fysiogr. Sälls. Lund Förhand. **29** (1959), 97–98. There is a tail-piece to show that the initial assumption that no body has actually *zero* mass can be dispensed with. This minute addition is not trivial; indeed it took a fortnight's damned hard work. (But though the essential idea takes pages to state it came to me on a walk, and this time literally in a *fraction* of a second.)

THE MATHEMATICIAN'S ART OF WORK

I will begin by saying, with a double motive, that there are a lot of queer people in the world. I remember a report of a man who was three times saved from drowning in a day, bathed once more and *was* drowned. The same year there was the case of a man, whom I think of as having spent his life in the British Museum, who conceived the idea of a seaside holiday, had himself rowed out, dived overboard, and being unable to swim, was drowned.

A former pupil began brilliantly; he took a pure research post after his Ph.D. under me, and had 6 years of research. The work became dull, though copious, and finally ended, and when I then met him he was on the point of a nervous breakdown. I then discovered that he had worked continuously for the 6 years for 365 and a quarter days a year. If he had done the work in the reverse order he would have been a Fellow of the Royal Society.

Shortly before World War I the psychologist, Boris Sidis, subjected his son to the theory of education. By the age of 9 the boy had become outstanding in a number of subjects. According to the theory no strain was involved, and in fact the boy did not obviously overwork and was also good at athletics. Some time after the war, when he was about 30, he was met by visitors from England. He held an ill-paid post with unexacting duties, refused to be promoted, and said that his object in life was never to have to *think* again.

These are awful examples. On the other hand, there are successfully creative people with strange methods of work. I know of a man who works only two days a week, of one who can work only in a cabaret, of one who has a wine-bottle by his desk. The economist, Marshall, though he had been through the mill of the Cambridge Mathematical Tripos and was second Wrangler, could not, or at any rate did not, work in later life for more than fifteen minutes at a stretch.

There are two morals to all this. If a young man feels he is not at home in the world, or that his instincts of how to work are abnormal, there is no reason for him to worry unduly. On the other hand he would be wise to find out what the usual methods are and give them a prolonged trial (less than a month is no good at all) before finally committing himself. There can be powerful illusions on such points, which I will come to later.

Creativity

At the lowest level there is an element of creativity in much ordinary conversation. We do not think what we are going to say and then say it, the experience is subjectively simple, and what is said emerges from the subconscious into the conscious. Long experience has established a working liason between the two, but if it fails, one becomes 'tongue-tied'.

At the other extreme of creativity, there are the 'great' creations, of something totally new and unexpected, and also of great importance, and seminal. We should all feel that the difference is one of kind and not of degree. (If I may frivolously digress, do you know the question: is the difference between a difference of degree and a difference of kind a difference of degree or a difference of kind? The *answer*, of course, is elementary.)

Much lies between the extremes. In any new form of mental activity, however man-made, the niche fills with people whose capacity is many orders of magnitude above the average. It has been said that we use only a small part of our brains; these facts are perhaps evidence for the idea. Let us run over some cases.

Oscar Wilde could cast down the pages of a novel, and in five minutes pass an examination on the contents.

The fantastic performances of musical prodigies and calculating boys are well-known. They have an intense interest in their game to the exclusion of all else, but the facts would be incredible if they did not happen. The calculators split into the above average intelligence and below. The former lose interest when they realize that anyone can get his results slowly by routine methods; they reach their peak at about the age of 4. Gauss was a case of this, though he did do a good deal of numerical calculation throughout his life: possibly he found in it a relaxation like that provided today by crosswords; but Gauss was a law to himself. A contemporary of Gauss, Däse, was of low intelligence, kept his capacity all his life, and was actually employed by Gauss to make factor-tables. Bidder, himself a highly intelligent case, had a daughter with a rather different faculty, but one out of all relation with ordinary people. She knew the current 707 digits for π, could begin at any point and read them off *either forwards or backwards*. She was studied by philosophers and psychologists in Cambridge, but she was quite unable to explain how she did it.

Computer theory has thrown up a class with suitable gifts, and they are not necessarily very good at mathematics. There is a class, not apparently very distinguished intellectually, which — as a recent experiment showed — can do difficult crossword puzzles with almost complete certainty and in an incredibly short time.

There are people who can learn a new language in a week, but most adults are poor linguists. Children, on the other hand, if suitably exposed, can be fluently trilingual by the age of six (more than three languages creates confusion). I don't know how to place this difference, but clearly this is the right way to be trilingual.

None of all this is highly creative. But, between the extremes, there is the army of people very gifted but short of genius. Though the importance of their creations falls short of the highest, I think the psychology involved is pretty much the same. A *sine qua non* is an intense conscious curiosity about the subject, with a craving to exercise the mind on it, quite like physical hunger. Love of truth — and all that — may co-exist, but I deny that it is the driving force. (To digress on physical analogies, a 'hunch' — an idea for which one can give no reason — seems analogous to smell.) Given the strong drive, it communicates itself in some form to the subconscious, which does all the real work, and would seem to be always on duty. Lacking the drive, one sticks. I have tried to learn mathematics outside my fields of interest; after any interval I had to begin all over again.

Four phases

It is usual to distinguish four phases in creation: preparation, incubation, illumination, and verification, or working out. For myself I regard the last as within the range of any competent practitioner, given the illumination.

Preparation is largely conscious, and anyhow *directed* by the conscious. The essential problem has to be stripped of accidentals and brought clearly into view; all relevant knowledge surveyed; possible analogues pondered. It should be kept constantly before the mind during intervals of other work. This last is advice from Newton.

Incubation is the work of the subconscious during the waiting time, which may be several years. Illumination, which can happen in a fraction of a second, is the emergence of the creative idea into the conscious. This almost always occurs when the mind is in a state of relaxation, and engaged lightly with ordinary matters. Helmholtz's ideas usually came to him when he was walking in hilly country. There is a lot to be said for walking during rest periods, unpopular as the idea may be. Incidentally, the relaxed activity of shaving can be a fruitful source of minor ideas; I used to postpone it, when possible, till after a period of work. Illumination implies some mysterious rapport between the subconscious and the conscious, otherwise emergence could not happen. What rings the bell at the right moment?

I recently had an odd and vivid experience. I had been struggling for two months to prove a result I was pretty sure was true. When I was walking up a Swiss mountain, fully occupied by the effort, a very odd device emerged — so odd that, though it worked, I could not grasp the resulting proof as a whole. But not only so; I had a sense that my subconscious was saying, 'Are you *never* going to do it, confound you; try this.'

My description so far has been appropriate to science or mathematics, one single idea. In the case of a symphony, incubation would be a more continuous process and many separate illuminations would be called for. And then the final miracle of a great symphony is the welding into an organic whole.

Beethoven's notebooks show that — with some remote objective vaguely in view — he would start with deliberate crudities, and approach the final work through blunders and repeated alterations. A recent surprise has been that the apparently completely spontaneous Dvořák was quite a similar case.

Some anecdotes

Kekulé's Benzene ring came in a dream. It is proverbial that sleep alters thoughts and decisions, but I believe high creation in dreams is very rare. William James seemed to have what seemed vitally important ideas in dreams, but always forgot them on waking. He decided to write down such dreams, and succeeded in doing so on the next occasion. In

... the relaxed activity of shaving can be a fruitful source of minor ideas; I used to postpone it, when possible, till after a period of work.

the morning he read 'Higamus hogamus, woman is monogamous; Hogamus higamus, man is polygamous.' It is not too bad: it has both form and content.

Mendeleev was a conscientious Professor of Chemistry with the urge to do the utmost for his pupils. He collected together likenesses of elements and tried to number them helpfully so as to be mnemonics. The final upshot was the Periodic Law.

Lobachevski had to teach Geometry; and began by an intensive critical study of Euclid, acting as devil's advocate. This approach to the anomalous parallel axiom resulted in his non-Euclidean Geometry.

M. Riesz had conceived a beautiful collection of theorems. He could prove them all if he could only prove a very special and innocent-looking case of one of them. *Specifically: if $f(z) = u + iv$ is regular in $|z| \leq 1$ and $f(0) = 0$ then we are to have

$$\int_{-\pi}^{\pi} |v|^4 d\theta \leq A \int_{-\pi}^{\pi} |u|^4 d\theta.$$

One day, having to give an examination, he was playing about with $\int_{-\pi}^{\pi} f^4 d\theta$ instead of the familiar $\int_{-\pi}^{\pi} |f|^4 d\theta$. This gives

$$\int_{-\pi}^{\pi} v^4 d\theta = -\int_{-\pi}^{\pi} u^4 d\theta + 6 \int_{-\pi}^{\pi} u^2 v^2 d\theta.$$

Using Cauchy's inequality on the least provocation was second nature to him. Doing this and changing the *minus* into a *plus* to fit higher indices, he had

$$\int_{-\pi}^{\pi} v^4 d\theta \leq \int_{-\pi}^{\pi} u^4 d\theta + 6\left(\int_{-\pi}^{\pi} u^4 d\theta \int_{-\pi}^{\pi} v^4 d\theta \right)^{1/2},$$

and saw his proof staring him in the face.*

A possible moral of the last three instances is that teaching activities may pay off in pure research.

In passing, I firmly believe that research should be offset by a certain amount of teaching, if only as a change from the agony of research. The trouble, however, I freely admit, is that in practice you get either no teaching, or else far too much.

Erasmus Darwin held that every so often you should try a damn-fool experiment. He played the trombone to his tulips. This particular result was, in fact, negative. But other incredibly impudent ideas have succeeded. An Italian physicist had rigged up two screens and transmitted electric impulses from one to the other. If he had spoken to one of the screens he would have invented the telephone, but with his sound physical sense he of course did nothing of the kind. The phonograph and telephone are surely the most impudent ideas.

There is a superb example, which I have described as the most impudent idea in Mathematics, with very important consequences; it is too technical for this paper, but you will find it discussed on pages 20–23 of my *Mathematician's Miscellany*. (Pages 40–43 of this present work).

Erasmus Darwin held that every so often you should try a damn-fool experiment. He played the trombone to his tulips.

On being a mathematician. There is much to be said for being a mathematician. To begin with, he has to be completely honest in his work, not from any superior morality, but because he simply cannot get away with a fake. It has been cruelly said of arts dons, especially in Oxford, that they believe there is a polemical answer to everything; nothing is really *true*, and in controversy the object is to prove your opponent a fool. We escape all this. Further, the arts man is always on duty as a great mind; if he drops a brick, as we say in England, it reverberates down the years. After an honest day's work a mathematician goes off duty. Mathematics is very hard work, and dons tend to be above average in health and vigor. Below a certain threshold a man cracks up; but above it, hard mental work *makes* for health and vigor (also — on much historical evidence throughout the ages — for longevity). I have noticed lately that when I am working really hard I wake around 5.30 a.m. ready and eager to start; if I am slack, I sleep till I am called. I mentioned this to a psychological doctor, who said it was now a known phenomenon.

There is one drawback to a mathematical life. The experimentalist, having spent the day looking for the leak, has had a complete mental rest. A mathematician's normal day contains hours of close concentration, and leaves him jaded in the evening. To appreciate something of high aesthetic quality needs close attention, easy to the unfatigued; but a strain for the fatigued mathematician. (Music seems a happy exception to this.) This is why we tend to relax either on mild nonfiction like biographies, or — to be crude, and to the derision of arts people — on trash. There is, of course, good trash and bad trash.

The higher mental activities are pretty tough and resilient, but it is a devastating experience if the drive does stop, and a long holiday is the only hope. Some people do lose it in their forties, and can only stop. In England they are a source of Vice-Chancellors.

Minor depressions will occur, and most of a mathematician's life is spent in frustration, punctuated with rare inspirations. A beginner can't expect quick results; if they are quick they are pretty sure to be poor.

To digress on this point, the ideal line for a supervisor with a really promising man is to give him two subjects: one, ambitious; the other, one that the supervisor can judge to be adequate for a Ph.D. (even if he has to do the thing himself first).

When one has finished a substantial paper there is commonly a mood in which it seems that there is really nothing in it. Do not worry, later on you will be thinking 'At least I could do something good *then*.' At the end of a particularly long and exacting work there can be a strange melancholy. This, however, is romantic, and mildly pleasant, like some other melancholies.

Research strategy

With a good deal of diffidence I will try to give some practical advice about research and the strategy it calls for. In the first place research work is of a different order from the 'learning' process of pre-research education (essential as it is). The latter can easily be rote-memory, with little associative power: on the other hand, after a month's immersion in research the mind knows its problem as much as the tongue knows the inside of one's mouth. You must also acquire the art of 'thinking vaguely,' an elusive idea I can't elaborate in short form. After what I have said earlier, it is inevitable that I should stress the importance of giving the subconscious every chance. There should be relaxed periods during the working day, profitably, I say, spent in walking.

Hours a day and days a week. On days free from research, and apart from regular holidays, I recommend four hours a day or at most five, with breaks about every hour (for walks perhaps). If you don't have breaks you unconsciously acquire the habit of slowing down. Preparation of lectures counts more or less as research work for this purpose. On days with teaching duties, I can only say, be careful not to overdo the research. The strain of lecturing, by the way, can be lightened if you apply the golfing maxim: 'Don't press.' It is, of course, hard not to. Don't spend tired periods on proof correction, or work that needs alertness; you make several shots at an emendation that you would do in one when fresh. Even in making a fair copy one is on the *qui vive* for possible changes.

Either work all out or rest completely. It is too easy, when rather tired, to fritter a whole day away with the intention of working but never getting properly down to it. This is pure waste, nothing is done, *and* you have had no rest or relaxation. I said 'Work all out': speed of associative thought is, I believe, important in creative work; another elusive idea, with which my psychological doctor agrees.

For a week without teaching duties — and here I think I am preaching to the converted — I believe in one afternoon and the following day off. The day off need not necessarily be Sunday, but that has a restful atmosphere of general relaxation, church bells in the distance, other people going to church, and so on. The day, however, should stay the same one of the week; this establishes a rhythm, and you begin relaxing at lunch time the day before.

At one time I used to work 7 days a week (apart, of course, from 3-week chunks of holidays). I experimented during the Long Vacation with a Sunday off, and presently began to notice that ideas had a way of coming on Mondays. I also planned to celebrate the arrival of a decent idea by taking the rest of that day off. And then ideas began coming also on Tuesday.

Morning versus evening. Before World War I it was usual in Cambridge to do our main work at night, 9.30 to 2.00 or later. Time goes rapidly — one has a whisky and soda at 11.30 and another later — and work *seems* to go well and easily. By comparison the morning seems bleak and work a greater effort. I am sure all this is one of the many powerful illusions about creative work. When put out of action by a severe concussion in 1918, I consulted Henry Head, an eminent psychologist, and known for wise hunches as a doctor. The traditional prescription was complete rest, but he told me to work as soon as I felt like it (I had leave of absence) and as much as I felt up to, *but* — only in the morning. After a month or two I discovered that, for me at least, morning work was far the better. I now never work after 6.30 p.m.

Warming up. Most people need half an hour or so before being able to concentrate fully. I once came across some wise advice on this, and have taken it. The natural impulse towards the end of a day's work is to finish the immediate job: this is of course right if stopping would mean doing work all over again. But try to end in the middle of something; in a job of writing out, stop in the middle of a sentence. The usual recipe for warming up is to run over the latter part of the previous day's work; this dodge is a further improvement.

Before coming to the subject of holidays I will say something about the various symptoms of overwork. I have wrongly disregarded them in the past; so doubtless others do too. One symptom can be muscular trouble. I once got into a vicious circle of feet painful enough to prevent exercise. I went to a masseuse who had the reputation of being a bit of a crank outside her work; she said my trouble was due to mental overwork. I am afraid I laughed, but I found she was perfectly right.

An ominous symptom is an obsession with the *importance* of work, and filling every moment to that end. The most infallible symptom is the anxiety dream. One struggles all night with a pseudo-problem — possibly with some odd relation to one's current job — and wakes in the morning quite unrefreshed.

Holidays. A governing principle is that 3 weeks, exactly 21 days — the period is curiously precise — is enough for recovery from the severest mental fatigue provided there is nothing pathological. This is expert opinion and my expertise agrees entirely, even to the point that, for example, 19 is not enough. Further, 3 weeks is more or less essential for much milder fatigue. So the prescription is 3 weeks holiday at the beginning of each vacation. It is vital, however, that it should be absolutely unbroken, whatever the temptation or provocation.

I believe there is a seasonal effect in mental activity, the trough coming about the middle or end of March. The academic year is surely involved, but there is probably a climatic factor. If this question is to be studied — and I think it should be — moments of high inspiration are far too rare and sporadic for statistics, but we could fall back on chess, bridge, and particularly crosswords, as tests of alertness and minor creation. My own evidence comes from a period of about 10 years when I played a form of two-pack patience, or solitaire, involving a very high element of skill. My curve of successes showed a general rise, with an unmistakable seasonal effect, a trough at the end of March. There was one anomaly, a March with a positive peak. I then recalled that I had taken a sabbatical term's leave from January through March, skiing in Switzerland.

Holiday activities. For many people these are highbrow: visits to Italian art galleries, a tour of Greece, and so on. I admire them, but do not share their tastes. I have an intense interest in music, but this does not need travel. I began skiing at 40 and rock-climbing at 43: these activities renewed my youth. They gave the most complete relaxation possible for mental work, and I believe they enlarge one's personality. Simple walking in mountains is very good. So is golf.

Food, tobacco, alcohol. Do not work within two hours of a substantial meal; blood cannot be in two places at once. I was once trapped into cold salmon at 6.30 p.m., immediately followed by a lecture, which I had to leave largely to the spur of the moment. The lecture was confused and I was poisoned for a week afterwards. I should have starved.

On tobacco my bleak advice is: no smoking till the day's work is over. There is much to be said against regular smoking: you are merely normal when smoking and miserable when not. I was converted from the heaviest possible smoking (16 pipes and 4 cigars a day) by Henry Head. He had the 1918 flu, followed by lack of tobacco on his ship and at his destination; an interval of 4 or 5 weeks. He decided to try giving it up for good. He then found that a heavy paper he was writing was finished in, he said, a third of the time it would have taken before. I sighed, succeeded in the struggle for abstinence, and I fully agreed with his estimate. I said speed was important; smoking is its enemy.

Alcohol is a depressant or sedative, not a stimulant, in spite of the illusion of champagne. It has a very valuable function in *stopping* thinking at the end of the day's work, a thing which many workers find indispensable. It has been said that, for this reason, Beethoven's posthumous quartets were paid for by cirrhosis of the liver. A permissible use is to mitigate the boredom of long routine calculations, or making a final draft. But then there must be a final check.

Tea and coffee. These are admirable, with no later reaction. They are usually considered to be rather mild in the way of stimulants; but since most of us never have the experience of being without them, it is possible that their virtue is underestimated.

. . . stimulants do exist; but they should be used . . . only in a crisis. And there is the problem of knowing what is a crisis.

Routine chores

It is a good idea to keep a set period in the week to deal with these. As a young man I found it almost impossible to answer letters, so perhaps this is true of others. It is fatal (as I found) to put them into a file marked 'Urgent'. The technique I recommend is: will this letter ultimately be answered? If yes, answer immediately or on the set period. If no, straight into the waste paper basket and forget it. This needs *knowing* oneself, a very important thing in many ways: not everyone does. There is a true story of an English clergyman offered a Colonial Bishopric. An inquiring visitor was told by the daughter of the house: 'Father is in the study praying for guidance, Mother is upstairs packing.'

Drugs. I can envisage a future in which stimulant drugs could raise mental activity for a set period of work, and relaxing ones give a suitable compensating period, perhaps of actual sleep. The present is a time of transition; stimulants do exist; but they should be used only with the greatest care and only in a crisis. And there is the problem of knowing what *is* a crisis.